PLUMBER'S AND PIPE FITTER'S CALCULATIONS MANUAL

R. Dodge Woodson

SECOND EDITION

McGRAW-HILL

New York   Chicago   San Francisco   Lisbon   London
Madrid   Mexico City   Milan   New Delhi   San Juan
Seoul   Singapore   Sydney   Toronto

The McGraw·Hill Companies

Cataloging-in-Publication Data is on file with the Library of Congress

2 3 4 5 6 7 8 9 0   DOC/DOC   0 1 0 9 8 7 6

ISBN 0-07-144868-3

The sponsoring editor for this book was Larry S. Hager and the production
supervisor was Pamela A. Pelton. It was set in Weiss by Lone Wolf Enterprises,
Ltd. The art director for the cover was Anthony Landi.

Printed and bound by RR Donnelley.

McGraw-Hill books are available at special quantity discounts to use as
premiums and sales promotions, or for use in corporate training programs. For
more information, please write to the Director of Special Sales, McGraw-Hill
Professional, Two Penn Plaza, New York, NY 10121-2298. Or contact your
local bookstore.

 This book is printed on recycled, acid-free paper containing a
minimum of 50% recycled, de-inked fiber.

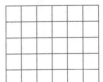

DEDICATION

I dedicate this book to Adam, Afton, and Victoria in appreciation for their patience during my writing time.

PREFACE

This book is your ticket to smooth sailing when it comes to doing the math for plumbing and pipe fitting. Most of the work is already done for you when you consult the many tables and references contained in these pages. Why waste time with calculators and complicated mathematical equations when you can turn to the ready-reference tables here and have the answers at your fingertips? There is no reason to take the difficult path when you can put your field skills to better use and make more money.

A few words of advice are needed here. Our country uses multiple plumbing codes. Every code jurisdiction can adopt a particular code and amend it to their local needs. It is impossible to provide one code source to serve every plumber's needs. The code tables in this book are meant to be used as representative samples of how to arrive at your local requirements, but they are not a substitution for your regional code book. Always consult your local code before installing plumbing.

The major codes at this time are the International Plumbing Code and the Uniform Plumbing Code. Both are excellent codes. There have been many code developments in recent years. In addition to these two major codes, there are smaller codes in place that are still active. I want to stress that this is not a handbook to the plumbing code; this is a calculations manual. If you are interested in a pure code interpretation, you can review one of my other Mc-Graw-Hill books entitled: *International and Uniform Plumbing Codes Handbook.*

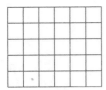

CONTENTS

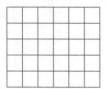

ABOUT THE AUTHOR

R. Dodge Woodson is a master plumber who lives in Maine and runs the plumbing, construction, and remodeling company The Masters Group, Inc. He has worked in the plumbing trade for 30 years and has written numerous books on plumbing. He has also been an instructor for the Central Maine Technical College for classes in code interpretation and apprenticeship.

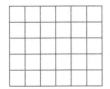

INTRODUCTION

Are you a plumber or pipe fitter who dislikes doing the math that is required in your trade? If so, this book may be one of the best tools that you can put in your truck or office. Why? Because it does much of the math calculations for you. That's right, the tables and visual graphics between these pages can make your life much easier and more profitable.

R. Dodge Woodson, the author, is a 30-year veteran of the trade who has been in business for himself since 1979. He knows what it takes to win in all financial climates as both a business owner and tradesman. This is your chance to learn from an experienced master plumber and, what is even better, you don't have to study and memorize formulas. All you have to do is turn to the section of this professional reference guide that affects your work and see the answers to your questions in black and white. How much easier could it be?

Mathematical matters are not the only treasures to be found here. You will find advice on how to comply with the plumbing code quickly, easily, and without as much thought on your part.

The backbone of this book is math for the trades, but there is much more. There is a section on troubleshooting that is sure to save you time, frustration, and money. Find out what you may need to know about septic systems. In addition to phase-specific math solutions, there is an appendix that is full of reference and conversion tables for day-to-day work situations.

Take a moment to scan the table of contents. You will see that the presentation of material here is compiled in logical, accessible, easy-to-use chapters. Flip through the pages and notice the tip boxes and visual nature of the information offered. You don't have to read much, but you will find answers to your questions.

If you are looking for a fast, easy, profitable way to avoid the dense reading and complicated math that is needed in your trade, you have found it. Once you put this ready reference guide at your fingertips, you will be able to concentrate on what you do best without the obstacles that may steal your time and your patience. Packed with 30 years of experience, you can't go wrong by using Woodson's resources to make you a better tradesman.

chapter 1

GENERAL TRADE MATHEMATICS

Math is not always a welcome topic among tradespeople. As much as math may be disliked, it plays a vital role in the trades, and plumbing and pipe fitting are no exceptions. In fact, the math requirements for some plumbing situations can be quite complicated. When people think of plumbers, few thoughts of scholarly types come to mind. I expect that most people would have trouble envisioning a plumber sitting at a drafting table and performing a variety of mathematical functions involving geometry, algebra, and related math skills. Yet, plumbers do use high-tech math in their trade, sometimes without realizing what they are doing.

Think about your last week at work. Did you work with degrees of angles? Of course you did. Every pipe fitting you installed was an example of angles. Did you grade your drainage pipe? Sure you did, and you used fractions to do it. The chances are good that you did a lot more math than you realized. But, can you find the volume of a water heater if the tank is not marked for capacity? How much water would it take to fill up a 4-inch pipe that is 100 feet long? You might need to know if you are hauling the water in for an inspection test of the pipe. How much math you use on a daily basis is hard to predict. Much of the answer would depend on the type of work you do within the trade. But, it's safe to say that you do use math on a daily basis.

I've taught a number of classes for plumbers and plumbing apprentices. Math is usually the least appreciated part of those classes. Experience has showed me that students resist the idea of learning math skills. I remember when I took academic levels of math in school and thought that I'd never use it. Little did I know back then how valuable the skills I was learning would be.

 been there **done that**

I was horrible with math in school. It was not until dollar signs were put in front of numbers that I understood math. When I entered the plumbing trade, I had no idea that I was doing a lot of math. If an employer had told me that math was a requirement for plumbers, I might not have devoted most of my adult life to the trade. Plumbing math doesn't seem like math, but it is serious math. Don't be afraid of it.

1

A or a	Area, acre
AWG	American Wire Gauge
B or b	Breadth
bbl	Barrels
bhp	Brake horsepower
BM	Board measure
Btu	British thermal units
BWG	Birmingham Wire Gauge
B & S	Brown and Sharpe Wire Gauge (American Wire Gauge)
C of g	Center of gravity
cond	Condensing
cu	Cubic
cyl	Cylinder
D or d	Depth, diameter
dr	Dram
evap	Evaporation
F	Coefficient of friction; Fahrenheit
F or f	Force, factor of safety
ft (or ′)	Foot
ft lb	Foot pound
fur	Furlong
gal	Gallon
gi	Gill
ha	Hectare
H or h	Height, head of water
HP	horsepower
IHP	Indicated horsepower
in (or ″)	Inch
L or l	Length
lb	Pound
lb/sq in.	Pounds per square inch
mi	Mile
o.d.	Outside diameter (pipes)
oz	Ounces
pt	Pint
P or p	Pressure, load
psi	Pounds per square inch
R or r	Radius
rpm	Revolutions per minute
sq ft	Square foot
sq in.	Square inch
sq yd	Square yard
T or t	Thickness, temperature
temp	Temperature
V or v	Velocity
vol	Volume
W or w	Weight
W. I.	Wrought iron

FIGURE 1.1 ■ **Abbreviations.** (*Courtesy of McGraw-Hill*)

While I'm not a rocket scientist, I can take care of myself when it comes to doing math for trade applications.

I assume that your time is valuable and that you are not interested in a college course in mathematics by the end of this chapter. We're on the same page of the playbook. I'm going to give you concise directions for solving mathematical problems that are related to plumbing and pipefitting. We won't be doing an in-depth study of the history of numbers, or anything like that. The work we do here will not be too difficult, but it will prepare you for the hurdles that you may have to clear as a thinking plumber. So, let's do it. The quicker we start, the quicker we can finish.

BENCHMARKS

Before we get into formulas and exercises, we need to establish some benchmarks for what we will be doing. It always helps to understand the terminology being used in any given situation, so refer to Figure 1.1 for reference to words and terms being used as we move forward in this chapter. The information in Figure 1.2 shows you some basic formulas that can be applied

---

Circumference of a circle = π × diameter or 3.1416 × diameter

Diameter of a circle = circumference × 0.31831

Area of a square = length × width

Area of a rectangle = length × width

Area of a parallelogram = base × perpendicular height

Area of a triangle = ½ base × perpendicular height

Area of a circle = π radius squared or diameter squared × 0.7854

Area of an ellipse = length × width × 0.7854

Volume of a cube or rectangular prism = length × width × height

Volume of a triangular prism = area of triangle × length

Volume of a sphere = diameter cubed × 0.5236 (diameter × diameter × diameter × 0.5236)

Volume of a cone = π × radius squared × ⅓ height

Volume of a cylinder = π × radius squared × height

Length of one side of a square × 1.128 = the diameter of an equal circle

Doubling the diameter of a pipe or cylinder increases its capacity 4 times

The pressure (in lb/sq in.) of a column of water = the height of the column (in feet) × 0.434

The capacity of a pipe or tank (in U.S. gallons) = the diameter squared (in inches) × the length (in inches) × 0.0034

1 gal water = 8½ lb = 231 cu in.

1 cu ft water = 62½ lb = 7½ gal

FIGURE 1.2 ■ Useful formulas. (*Courtesy of McGraw-Hill*)

Sine	$\sin = \dfrac{\text{side opposite}}{\text{hypotenuse}}$
Cosine	$\cos = \dfrac{\text{side adjacent}}{\text{hypotenuse}}$
Tangent	$\tan = \dfrac{\text{side opposite}}{\text{side adjacent}}$
Cosecant	$\csc = \dfrac{\text{hypotenuse}}{\text{side opposite}}$
Secant	$\sec = \dfrac{\text{hypotenuse}}{\text{side adjacent}}$
Cotangent	$\cot = \dfrac{\text{side adjacent}}{\text{side opposite}}$

FIGURE 1.3 ■ Trigonometry. (*Courtesy of McGraw-Hill*)

▶ *sensible* **shortcut**

You don't have to do the math if you have reliable tables to use when arriving at a viable answer for mathematical questions. The types of tables that you need to limit your math requirements are available in this book.

to many mathematical situations. Trigonometry is a form of math that can send some people in the opposite direction. Don't run, it's not that bad. Figure 1.3 provides you with some basics for trigonometry, and Figure 1.4 describes the names of shapes that contain a variety of sides. Some more useful formulas are provided for you in Figure 1.5. Just in what I've provided here, you are in a much better position to solve mathematical problems. But, you probably want, or need, a little more explanation of how to use your newfound resources. Well, let's do some math and see what happens.

Pentagon	5 sides
Hexagon	6 sides
Heptagon	7 sides
Octagon	8 sides
Nonagon	9 sides
Decagon	10 sides

FIGURE 1.4 ■ Polygons. (*Courtesy of McGraw-Hill*)

Parallelogram	Area = base × distance between the two parallel sides
Pyramid	Area = ½ perimeter of base × slant height + area of base
	Volume = area of base × ⅓ of the altitude
Rectangle	Area = length × width
Rectangular prism	Volume = width × height × length
Sphere	Area of surface = diameter × diameter × 3.1416
	Side of inscribed cube = radius × 1.547
	Volume = diameter × diameter × diameter × 0.5236
Square	Area = length × width
Triangle	Area = one-half of height times base
Trapezoid	Area = one-half of the sum of the parallel sides × the height
Cone	Area of surface = one-half of circumference of base × slant height + area of base
	Volume = diameter × diameter × 0.7854 × one-third of the altitude
Cube	Volume = width × height × length
Ellipse	Area = short diameter × long diameter × 0.7854
Cylinder	Area of surface = diameter × 3.1416 × length + area of the two bases
	Area of base = diameter × diameter × 0.7854
	Area of base = volume ÷ length
	Length = volume ÷ area of base
	Volume = length × area of base
	Capacity in gallons = volume in inches ÷ 231
	Capacity of gallons = diameter × diameter × length × 0.0034
	Capacity in gallons = volume in feet × 7.48
Circle	Circumference = diameter × 3.1416
	Circumference = radius × 6.2832
	Diameter = radius × 2
	Diameter = square root of = (area ÷ 0.7854)
	Diameter = square root of area × 1.1233

FIGURE 1.5 ■ Area and other formulas. (*Courtesy of McGraw-Hill*)

PIPING MATH

This section will profile formulas that can help you when working with pipes. Rather than talk about them, let's look at them.

What plumber hasn't had to figure the grading for a drainage pipe? Determining the amount of fall needed for a drainpipe over a specified distance is no big mystery. Yet, I've known good plumbers who had trouble with calculating the grade of their pipes. In fact, some of them were so unsure of themselves that they started at the end of their runs and worked backwards, to the beginning, to insure enough grade. Not only is this more difficult and time consuming, there is still no guarantee that there will be enough room for the grade. Knowing how to figure the grade, fall, pitch, or whatever you want to call it, for a pipe is essential in the plumbing trade. And, it's not difficult. Let me show you what I mean.

In a simple way of putting it, assume that you are installing a pipe that is 20 feet long and that will have a grade of ¼-inch per foot. What will the drop from the top of the pipe be from one end to the other? At a grade of ¼-inch per foot, the pipe will drop one inch for every four feet it travels. A 20-foot piece of pipe will require a 5-inch drop in the scenario described. By dividing 4 into 20, I got an answer of 5, which is the number of inches of drop. That's my simple way of doing it,

> ☑ *fast code* **fact**
>
> As a rule of thumb, most codes require a minimum of one-quarter of an inch per foot of fall for drainage piping. There are exceptions. For example, large-diameter pipes may be installed with a minimum grade of one-eighth of an inch per foot. Too much grade is as bad as too little grade. A pipe with excessive grade will empty liquids before solids have cleared the pipe. Maintain a constant grade within the confines of your local plumbing code.

The capacity of pipes is as the square of their diameters. Thus, doubling the diameter of a pipe increases its capacity four times. The area of a pipe wall may be determined by the following formula:

$$\text{Area of pipe wall} = 0.7854 \times [(\text{o.d.} \times \text{o.d.}) - (\text{i.d.} \times \text{i.d.})]$$

FIGURE 1.6 ■ Piping. (*Courtesy of McGraw-Hill*)

The approximate weight of a piece of pipe may be determined by the following formulas:

Cast-iron pipe: weight $= (A^2 - B^2) \times \text{length} \times 0.2042$
Steel pipe: weight $= (A^2 - B^2) \times \text{length} \times 0.2199$
Copper pipe: weight $= (A^2 - B^2) \times \text{length} \times 0.2537$
A = outside diameter of the pipe in inches
B = inside diameter of the pipe in inches

FIGURE 1.7 ■ Determining pipe weight. (*Courtesy of McGraw-Hill*)

The formula for calculating expansion or contraction in plastic piping is:

$$L = Y \times \frac{T - F}{10} \times \frac{L}{100}$$

L = expansion in inches
Y = constant factor expressing inches of expansion per 100°F temperature change
    per 100 ft of pipe
T = maximum temperature (°F)
F = minimum temperature (°F)
L = length of pipe run in feet

FIGURE 1.8 ▪ Expansion in plastic piping. (*Courtesy of McGraw-Hill*)

The formulas for pipe radiation of heat are as follows:

$$L = \frac{144}{OD \times 3.1416} \times R \div 12$$

D = outside diameter (OD) of pipe
L = length of pipe needed in feet
R = square feet of radiation needed

FIGURE 1.9 ▪ Formulas for pipe radiation of heat. (*Courtesy of McGraw-Hill*)

but now let me give you the more proper way of doing it with a more sophisticated formula.

If you are going to use the math formula, you must know the terms associated with it. Run is the horizontal distance that the pipe you are working with will cover, and this measurement is shown as the letter R. Grade is the slope of the pipe and is figured in inches per foot. To define grade in a formula, the letter G is used. Drop is the amount down from level or in more plumber-friendly words, it's the difference in height from one end of the pipe to the other. As you might guess, drop is known by the letter D. Now let's put this into a formula. To determine grade with the formula above, you would be looking at something like this: D = G × R. If you know some of the variables, you can find the rest. For example, if you know how far the pipe has to run and what the maximum amount of drop can be, you can determine the grade. When you know the grade and the length of the run, you can determine the drop. I already showed you how to find the drop if you know grade and run numbers. So, let's assume an example where you know that the drop is 15 inches and the run is 60 feet, what is the grade? To find the answer, you divide the drop by the run, in this case you are dividing 15 by 60. The answer is .25 or ¼-inch per foot of grade.

TEMPERATURE TIPS

Let me give you a few illustrations here that will help you deal with temperatures, heat loss, and mixing temperatures.

Temperature may be expressed according to the Fahrenheit (F) scale or the Celsius (C) scale. To convert °C to °F or °F to °C, use the following formulas:

$$°F = 1.8 \times °C + 32$$
$$°C = 0.55555555 \times °F - 32$$
$$°C = °F - 32 \div 1.8$$
$$°F = °C. \times 1.8 + 32$$

FIGURE 1.10 ■ Temperature conversion. (*Courtesy of McGraw-Hill*)

To figure the final temperature when two different temperatures of water are mixed together, use the following formula:

$$\frac{(A \times C) + (B \times D)}{A + B}$$

A = weight of lower temperature water
B = weight of higher temperature water
C = lower temperature
D = higher temperature

FIGURE 1.11 ■ Computing water temperature. (*Courtesy of McGraw-Hill*)

Radiation

3 ft of 1-in. pipe equal 1 ft² R.
2⅓ lineal ft of 1¼-in. pipe equal 1 ft² R.
Hot water radiation gives off 150 Btu/ft² R/hr.
Steam radiation gives off 240 Btu/ft² R/hr.
On greenhouse heating, figure ⅔ ft² R/ft² glass.
1 ft² of direct radiation condenses 0.25 lb water/hr.

FIGURE 1.12 ■ Radiant heat facts. (*Courtesy of McGraw-Hill*)

−100°–30°		
°C	Base temperature	°F
−73	−100	−148
−68	−90	−130
−62	−80	−112
−57	−70	−94
−51	−60	−76
−46	−50	−58
−40	−40	−40
−34.4	−30	−22
−28.9	−20	−4
−23.3	−10	14
−17.8	0	32
−17.2	1	33.8
−16.7	2	35.6
−16.1	3	37.4
−15.6	4	39.2
−15.0	5	41.0
−14.4	6	42.8
−13.9	7	44.6
−13.3	8	46.4
−12.8	9	48.2
−12.2	10	50.0
−11.7	11	51.8
−11.1	12	53.6
−10.6	13	55.4
−10.0	14	57.2
31°–71°		
°C	Base temperature	°F
−0.6	31	87.8
0	32	89.6
0.6	33	91.4
1.1	34	93.2
1.7	35	95.0
2.2	36	96.8
2.8	37	98.6
3.3	38	100.4
3.9	39	102.2
4.4	40	104.0
5.0	41	105.8
5.6	42	107.6

FIGURE 1.13 ■ Temperature conversion. (*Courtesy of McGraw-Hill*)

Vacuum in inches of mercury	Boiling point
29	76.62
28	99.93
27	114.22
26	124.77
25	133.22
24	140.31
23	146.45
22	151.87
21	156.75
20	161.19
19	165.24
18	169.00
17	172.51
16	175.80
15	178.91
14	181.82
13	184.61
12	187.21
11	189.75
10	192.19
9	194.50
8	196.73
7	198.87
6	200.96
5	202.25
4	204.85
3	206.70
2	208.50
1	210.25

FIGURE 1.14 ■ Boiling points of water based on pressure. (*Courtesy of McGraw-Hill*)

HOW MANY GALLONS?

How many gallons does that tank hold? Do you know how to determine the capacity of a tank? Well if you don't, you're about to see an easy way to find out. Before you can start to do your math, you have to know if you will be working with measurements in inches or in feet. You also have to know that the tank diameter is known as D and the tank height is H. We are looking for the tank capacity in gallons, which we will identify in our formula with the letter G.

When the measurements for a tank are expressed in inches, you will use a factor of 0.0034 in your formula. Tanks that are measured in terms of feet require a factor of 7.5. For our example, we are going to measure our tank in inches. This particular tank is 18 inches in diameter and 60 inches in height. The generic formula for this type of problem is as follows: $G = d^2 \times h \times 0.0034$. We know some of the variables, so we have to put them into our equation.

The diameter of our tank is 18 inches and the height is 60 inches, so our formula will look like this: G = 18^2 × 60 × 0.0034. What is 18^2? It's 324. This is found by multiplying 18 by itself or 18 × 18. Now we know that we are going to multiply 324 by 60 as we follow our formula. This will give us a number of 19440. The last step of our formula is to multiply 19440 by the 0.0034 factor. This will result in an answer of 66.10. We are looking for the maximum capacity of the tank, so we adjust the 66.10 to an even 66 gallons. That wasn't too bad, was it?

CYLINDER-SHAPED CONTAINERS

Cylinder-shaped containers could be tanks, pipes, or any other number of devices. What happens if you want to know the holding capacity of such an object? You are going to need to use a formula that involves the radius (R) of the object, the diameter of the object (D), the height (H) of the object, and the value assigned to π, which is 3.1416. Our goal is to find the volume (V) capacity of a cylinder. There are two types of formulas that can be used to determine the capacity of a cylinder, so let's take them one at a time.

The first formula that we are going to use looks like this: V = πr^2 h. Another way to find the answer is to have V = π divided by 4 d^2 h. Either formula will give you the same answer, it's just a matter of choosing one formula over another, based on your known elements of the question.

A LITTLE GEOMETRY

A little geometry is needed in the plumbing trade. Whether you are working with roof drains, figuring floor drains, or doing almost any part of plumbing paperwork, you may be using geometry. I hated geometry in school, but I've learned how to use it in my trade and how to make the use of it much more simple than I ever used to know it to be. I'll share some of my secrets on the subject.

Plumbers use geometry to find the distance around objects, to find the area of objects, to determine volume capacities, and so forth. A lot of plumbers probably don't think about what they are doing as geometry, but it is. So, let me show you some fast ways to solve your on-the-job problems by using geometry that you may not even realize is geometry. Think of what we are about to do as just good old plumbing stuff that has to be done.

Rectangles

Rectangles are squares, right? Wrong, they are rectangles. Squares are squares. Got ya! Now that I have your attention, let's talk about the methods used to determine perimeter measurements for a rectangle. A flat roof on a commercial building is a good example of a rectangle that a plumber might need to work with for rainwater drainage. This exercise is too simple. To find the perimeter (P), you multiply the length (L) by 2 and add it to the width (W) that has also been multiplied by two. The formula looks like this: P = 2L × 2W. Now let's put this into real numbers. Assume that you have a roof that is 80 feet long and 40 feet wide. What is the perimeter of the roof? First, do the math for the length. Taking 80 × 2 will give you 160. Do the width next. You will find that 40 × 2 is 80. When you add the 80 to the 160, you get 240, which is the perimeter of the roof. Not too tough, huh? Didn't I tell you that I'd make this stuff easy?

A Square

A square has a perimeter measurement. Do you know how to find it? This one really is too simple. Add up the measurements of the four equal sides and you have the perimeter. In other words, if you are dealing with a flat roof that is square with dimensions of 50 feet on each side, the perimeter is 200 feet. This is established by multiplying 50 × 4. They don't get any easier than this one.

Triangle Perimeters

Triangle perimeters are not difficult to establish. The process is similar to the one used with squares, only there is one less measurement. To find the perimeter of a triangle, add up the sum total of the three sides of the shape. If you want a formula to use, it could look like this: P = A + B + C. The long and the short of it, no pun intended, is that you simply add up the three dimensions and you have the perimeter.

Circles

Circles can give you some trouble when you are looking for their perimeters, which should really be called their circumference. I have provided resource tables in the next chapter that will help you to avoid doing the math to find the circumference of a circle, but we should at least take a few moments to

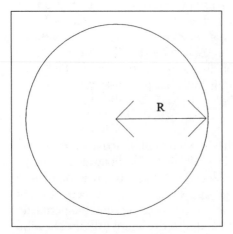

FIGURE 1.15 ■ Radius of a circle.

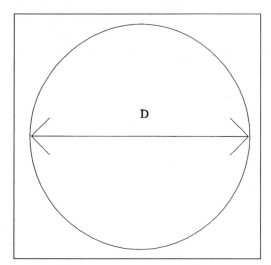

FIGURE 1.16 ■ Diameter of a circle.

explore the procedure while we are here. Circles can be tricky, but they aren't really all that tough. Let's look at a couple of formulas that you shouldn't experience problems with (Fig. 1.16).

When you want to find the circumference of a circle, you must work with the diameter (D), the radius (R), and π, which is 3.1416. You can use one of two formulas to solve your problem, depending on which variable is known. If you know the diameter, use the following formula: $C = \pi d$. When you know the radius, use this formula: $C = 2\pi r$. If the diameter is six inches, your formula would reveal that pi (3.1416) times 6 inches equals 18.8496 inches. This number would be rounded to 18.85 inches. If you knew the radius and not the diameter, your numbers would be 2 times π (3.1416) times 3 inches. The same answer would be arrived at, for a circumference of 18.85 inches. The formulas are not difficult, but using the tables in the next chapter might be faster and easier for you.

FINDING THE AREA AND VOLUME OF A GIVEN SHAPE

Finding the area of a given shape is also done with the use of formulas. It's no more difficult than what we have already been doing. In some ways, finding the area is easier than finding the perimeter. Most anyone in the trades knows how to find the square footage of a room. When you multiply the length of the room by the width of the room, you arrive at the square footage (Fig. 1.17). Well, this is exactly how you find the area of a rectangle or a square. There is no mystery or trick. Just multiply the length by the width for a rectangle or multiply one side by another side for a square, and you will have the area of the shape. To find the volume of a rectangle, you simply multiply the length by the width by the height. Different formulas are needed to find the area of trapezoids and triangles (Fig. 1.18 and

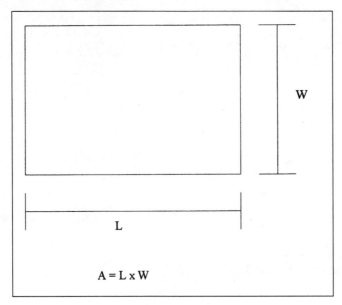

FIGURE 1.17 ■ Area of a rectangle.

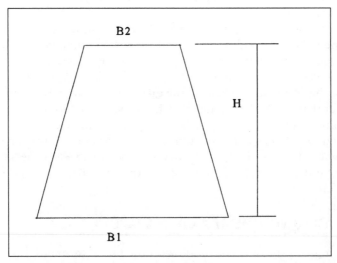

FIGURE 1.18 ■ Area of a trapezoid.

Fig. 1.19). A triangular prism requires yet a different formula when the volume of the shape is being sought (Fig. 1.20).

Want to find the area of a circle? The area will be equal to π (3.1416) multiplied by the radius squared. If we say that the radius of a circle is nine inches, we would start to find the area of the circle by multiplying 3.1416

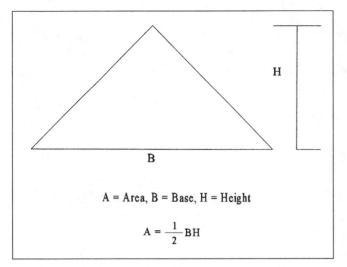

FIGURE 1.19 ■ Area of a triangle.

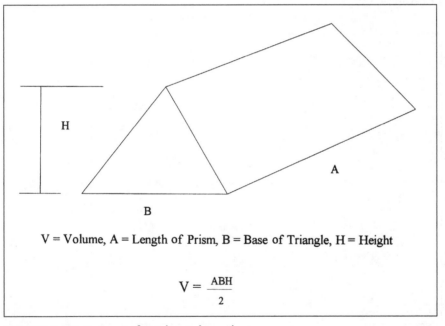

FIGURE 1.20 ■ Area of a triangular prism.

(π) by 9 inches by 9 inches. This would advance up to multiplying 3.1416 by 81 square inches ($9 \times 9 = 81$). The area of the circle would turn out to be 254.47 square inches. If you are looking for the volume of a cube, you simply multiply the three sides, as is illustrated in Figure 1.21. For a trapezoidal prism, the volume is found by using the formula in Figure 1.22.

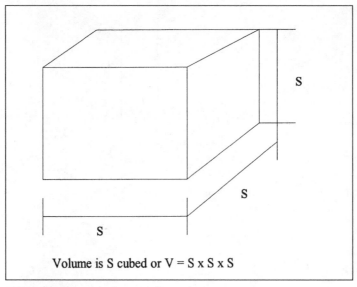

Volume is S cubed or V = S x S x S

FIGURE 1.21 ■ Volume of a cube.

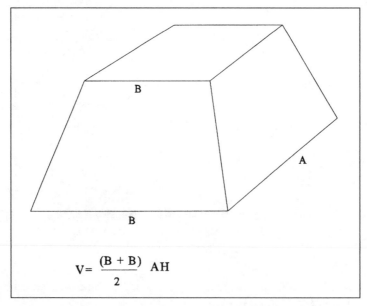

$$V = \frac{(B + B)}{2} AH$$

FIGURE 1.22 ■ Volume of a trapeziodal prism.

The math that is used in plumbing and pipe fitting is not very difficult to understand if you will accept the fact that it is necessary and that you need to understand it. What may appear daunting on the surface is actually pretty practical in principle. With a combination of reference tables, a good calculator, and a little effort, you can accomplish your needs for math within the trade quickly.

chapter 2

FORMULAS FOR PIPE FITTERS

Plumbing and pipe fitting are similar, but not always the same. Modern plumbers usually work with copper tubing and various forms of plastic piping. Cast-iron pipe is still encountered, and steel pipe is used for gas work. Finding a plumber working with threaded joints is not nearly as common as it once was. But, threaded pipe is still used in plumbing, and it is used frequently in pipe fitting. Figuring the fit for a pipe where threads are to be inserted into a fitting is a little different from sliding copper or plastic pipe into a hub fitting. However, many of the calculations used with threaded pipe apply to other types of pipe.

Many plumbers don't spend a lot of time using mathematical functions to figure offsets. Heck, I'm one of them. How often have you taken a forty-five and held it out to guesstimate a length for a piece of pipe? If you have a lot of experience, your trained eye and skill probably gave you a measurement that was close enough for plastic pipe or copper tubing. I assume this because I do it all the time. But, there are times when it helps to know how to use a precise formula to get an accurate measurement. The need for accuracy is more important when installing threaded pipe. For example, you can't afford to guess at a piece of gas pipe and find out the hard way that the threads did not go far enough into the receiving fitting.

In the old days, when I was first learning the trade, plumbers taught their helpers and apprentices. Those were the good-ole days. In today's competitive market, plumbing companies don't spend nearly as much time or money training their up-and-coming plumbers. As the owner of a plumbing company, I understand why this is, but I don't agree with it. And, the net result is a crop

> ▶ *sensible* **shortcut**
>
> When you test gas piping for leaks, you can use soapy water or a spray window cleaner to find the leak. Wipe or spray the solution on pipe threads and watch for bubbles to form. If they do, you have found the leak.

of plumbers who are not well prepared for what their trade requires. Sure, they can do the basics of gluing, soldering, and simple layouts, but many of the new breed don't possess the knowledge needed to be true master plumbers. Don't get me wrong; it's not really the fault of the new plumbers. Responsibility for becoming an excellent plumber rests on many shoulders.

Ideally, plumbing apprentices and helpers should have classroom training. Company supervisors should authorize field plumbers some additional time for in-the-field training for apprentices. Working apprentices should go the extra mile to do research and study on their own. When I was helper, I used to spend my lunch break reading the codebook. There is no single individual to blame for the quality of education that some new plumbers are, or are not, receiving. Money is probably the root of the problem. Customers are looking for low bids. Contractors must be competitive, and this eliminates the ability to have a solid on-the-job training program. Many helpers today seem to be more interested in getting their check than getting an education. So, here we are, with a lot of plumbers who don't know the inner workings of the finer points of plumbing.

☞ been there **done that**

If you are entering the plumbing trade, shop your-self to companies who will train you. I turned down jobs that offered more money than other companies when I was a helper. Why did I do this? Because I wanted to be a plumber. It is often the companies who pay less who will invest in training a rookie. The knowledge you gain will be worth much more than the extra money that you might make at another company. The money is likely to be gone in a year, but the knowledge will be with you for life.

▶ *sensible* **shortcut**

There is no sensible shortcut for learning your trade. It's okay to be on the earn-while-you-learn plan, but you have to apply yourself to become a true professional in the plumbing trade.

I was fortunate enough to be what might have been the last generation of plumbers to get company support in learning the trade. Plenty of time was spent running jackhammers and using shovels, but my field plumber took the time to explain procedures to me. I learned quickly how to plumb a basic house. Then I learned how to run gas pipe and to do commercial buildings. As a part of my learning process, I read voraciously. Later I became a supervisor, then the owner of my own company, and eventually an educator for other plumbers and for apprentices. I could have stopped anywhere along the way, but I've taken my interest in the trade to the limits, and I continue to push ahead. No, I don't know all there is to know, but I've worked hard to gain the knowledge I have. Now is the time for me to share my knowledge of pipe fitting math with you.

45° OFFSETS

Offsets for 45° bends are common needs in both plumbing and pipe fitting. In fact, this degree of offset is one of the most common in the trade. I mentioned

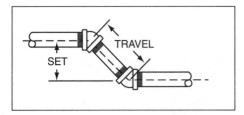

FIGURE 2.1 ■ Calculated 45° offsets.

earlier that many plumbers eyeball such offset measurements. The method works for a lot of plumbers, but let's take a little time to see how the math of such offsets can help you in your career.

To start our tutorial, let's discuss terms that apply to offsets. Envision a horizontal pipe that you want to install a 45° offset in. For the ease of vision, think of the horizontal pipe resting in a pipe hanger. You have to offset the pipe over a piece of ductwork. This will have a 45° fitting looking up from your horizontal pipe. There will be a piece of pipe in the upturned end of the fitting that will come into the bottom of the second 45° fitting (Fig. 2.1: offset drawing).

As we talk about measurements here, they will all be measured from the center of the pipe. There are two terms you need to know for this calculation. Travel, the first term, is the length of the pipe between the two 45° fittings. The length of Travel begins and ends at the center of each fitting. The distance from the center of the lower horizontal pipe to the center of the upper horizontal pipe is called the Set. Now that you know the terms, we can do the math.

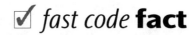

☑ *fast code* **fact**

Code measurements are typically based on measurements made from the centerline of fittings and pipe.

To make doing the math easier, I am including tables for you to work from (Fig. 2.2: 45° offset math tables). Let's say that the Set is $56\frac{3}{4}$ inches. Find this measurement in the table in Fig. 2.2. This will show you that the Travel is 80.244 inches. Now you can use the table for converting decimal equivalents of fractions of an inch (Fig. 2.3: decimal equivalents of fractions of an inch) to convert your decimal, the 80.244 inches. Finding the decimal equivalent of a fraction is a matter of dividing the numerator by the denominator. The chart in Fig. 2.3 proves the measurement to be $80\frac{1}{4}$ inches. You can find the Set if you know the Travel by reversing the procedure.

If the Travel is known to be $80\frac{1}{4}$ inches, what is the Set? We both know that it is $56\frac{3}{4}$ inches, but how would you find it? Use the table in Fig. 2.2 and look under the heading of Travel. Find the 80.244 listing that represents $80\frac{1}{4}$ inches. Refer to the Set heading. What does it say? Of course, it says $56\frac{3}{4}$. It's that easy. All you have to do is use the tables that I've provided to make your life easier in calculating 45° offsets.

Set	Travel	Set	Travel	Set	Travel
2	2.828	¼	15.907	½	28.987
¼	3.181	½	16.261	¾	29.340
½	3.531	¾	16.614	21	29.694
¾	3.888	12	16.968	¼	30.047
3	4.242	¼	17.321	½	30.401
¼	4.575	½	17.675	¾	30.754
½	4.949	¾	18.028	22	31.108
¾	5.302	13	18.382	¼	31.461
4	5.656	¼	18.735	½	31.815
¼	6.009	½	19.089	¾	32.168
½	6.363	¾	19.442	23	32.522
¾	6.716	14	19.796	¼	32.875
5	7.070	¼	20.149	½	33.229
¼	7.423	½	20.503	¾	33.582
½	7.777	¾	20.856	24	33.936
¾	8.130	15	21.210	¼	34.289
6	8.484	¼	21.563	½	34.643
¼	8.837	½	21.917	¾	34.996
½	9.191	¾	22.270	25	35.350
¾	9.544	16	22.624	¼	35.703
7	9.898	¼	22.977	½	36.057
¼	10.251	½	23.331	¾	36.410
½	10.605	¾	23.684	26	36.764
¾	10.958	17	24.038	¼	37.117
8	11.312	¼	24.391	½	37.471
¼	11.665	½	24.745	¾	37.824
½	12.019	¾	25.098	27	38.178
¾	12.372	18	25.452	¼	38.531
9	12.726	¼	25.805	½	38.885
¼	13.079	½	26.159	¾	39.238
½	13.433	¾	26.512	28	39.592
¾	13.786	19	26.866	¼	39.945
10	14.140	¼	27.219	½	40.299
¼	14.493	½	27.573	¾	40.652
½	14.847	¾	27.926	29	41.006
¾	15.200	20	28.280	¼	41.359
11	15.554	¼	28.635	½	41.713

FIGURE 2.2 ■ Set and travel relationships in inches for 45° offsets.

BASIC OFFSETS

Basic offsets are all based on the use of right triangles. You now know about Set and Travel. It is time that you learned about a term known as Run. Travel, as I said earlier, is the distance between center of two offset fittings that creates the length of a piece of pipe. This pipe's length is determined as it develops from fitting to fitting, traveling along the angle of the offset. When you want to know the Run, you are interested in the distance measured along a straight line from the bottom horizontal pipe. Refer to Fig. 2.4 for an example

Inches	Decimal of an inch	Inches	Decimal of an inch
1/64	.015625	33/64	.515625
1/32	.03125	17/32	.53125
3/64	.046875	35/64	.546875
1/16	.0625	9/16	.5625
5/64	.078125	37/64	.578125
3/32	.09375	19/32	.59375
7/64	.109375	39/64	.609375
1/8	.125	5/8	.625
9/64	.140625	41/64	.640625
5/32	.15625	21/32	.65625
11/64	.171875	43/64	.671875
3/16	.1875	11/16	.6875
13/64	.203125	45/64	.703125
7/32	.21875	23/32	.71875
15/64	.234375	47/64	.734375
1/4	.25	3/4	.75
17/64	.265625	49/64	.765625
9/32	.28125	25/32	.78125
19/64	.296875	51/64	.796875
5/16	.3125	13/16	.8125
21/64	.328125	53/64	.828125
11/32	.34375	27/32	.84375
23/64	.359375	55/64	.859375
3/8	.375	7/8	.875
35/64	.390625	57/64	.890625
13/32	.40625	22/32	.90625
27/64	.421875	59/64	.921875
7/16	.4375	15/16	.9375
29/64	.453125	61/64	.953125
15/32	.46875	31/32	.96875
31/64	.484375	63/64	.984375
1/2	.5	1	1

FIGURE 2.3 ▪ Decimal equivalents of fractions of an inch.

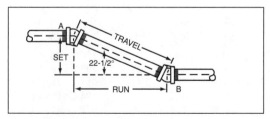

FIGURE 2.4 ▪ Simple offsets.

of what I'm talking about. Run is a term applied to the horizontal measurement from the center of one offset fitting to the center of the other offset fitting.

Most charts and tables assign letters to terms used in formulas. For our purposes, let's establish our own symbols. We will call the letter S–Set, the letter R–Run, and the letter T–Travel. What are common offsets in the plumbing and pipe fitting trade? A 45° offset is the most common. Two other offsets sometimes use are 60° bends and 22½° bends. These are the three most frequently used offsets and the ones that we will concentrate our efforts on.

The use of the right triangle is important when dealing with piping off-sets. The combination of Set, Travel, and Run form the triangle. I can provide you with a table that will make calculating offsets easier (Fig. 2.5), but you must still do some of the math yourself, or at least know some of the existing figures. This may seem a bit intimidating, but it is not as bad as you might think. Let me explain.

As a working plumber or pipe fitter, you know where your first pipe is. In our example earlier, where there was ductwork that needed to be cleared, you can easily determine what the measurement of the higher pipe must be. This might be determined by measuring the distance from a floor or ceiling. Either way, you will know the center measurement of your existing pipe and the cen-ter measurement for where you want the offset pipe to comply with. Know-ing these two numbers will give you the Set figure. Remember, Set is meas-ured as the vertical distance between the centers of two pipes. Refer back to Fig. 2.1 if you need a reminder on this concept.

Let's assume that you know what your Set distance is. You want to know what the Travel is. To do this, use the table in 2.5. For example, if you were looking for the Travel of a 45° offset when the Set is known, you would mul-tiply the Set measurement by a factor of 1.414. Now, let's assume that you know the Travel and want to know the Set. For the same 45° offset, you would multiply the Travel measurement by .707. It's really simple, as long you have the chart to use. The procedure is the same for different degrees of offset. Just refer to the chart and you will find your answers quickly and easily.

To find side*	When known side is	Multi- ply Side	For 60° ells by	For 45° ells by	For 30° ells by	For 22½° ells by	For 11¼° ells by	For 5⅝° ells by
T	S	S	1.155	1.414	2.000	2.613	5.125	10.187
S	T	T	.866	.707	.500	.383	.195	.098
R	S	S	.577	1.000	1.732	2.414	5.027	10.158
S	R	R	1.732	1.000	.577	.414	.198	.098
T	R	R	2.000	1.414	1.155	1.082	1.019	1.004
R	T	T	.500	.707	.866	.924	.980	.995

*S = set, R = run, T = travel.

FIGURE 2.5 ■ **Multipliers for calculating simple offsets.**

Finding Run measurements is no more difficult than Set or Travel. Say you have the Set measurement and want to know the Run figure for a 45° offset. Multiply the Set figure by 1.000 to get the Run number. If you are working with the Travel number, multiply that number by .707 to get the Run number for a 45° offset.

SPREADING OFFSETS EQUALLY

If you take a lot of pride in your work or are working to detailed piping diagrams, you may find that the spacing of your offsets must be equal. Equally-spaced offsets are not only more attractive and more professional looking, they might required. You can guess and eyeball measurements to get them close, but you will need a formula to work with if you want the offsets to be accurate. Fortunately, I can provide you with such a formula, and I will.

Again, we will concentrate on 45°, 60°, and 22½° bends, since these are the three most often used in plumbing and pipefitting. We will start with the 45° turns. In our example, you should envision two pipes rising vertically. Each pipe will be offset to the left and then the pipes will continue to rise vertically. For a visual example, refer to Fig. 2.6. It is necessary for us to determine uniform symbols for what we are doing, so let's get that out of the way right now.

In our measurement examples, we will refer to Spread, the distance between the two offsetting pipes from center to center, as A. Set will remain with the symbol of S. Travel will be T and it will be the same as Distance of D. Run will be noted by the letter R. The letter F will be the length of pipe threads.

Now for the deal. Travel is determined in an equally-offset pipe run at a 45° angle by multiplying the Set by 1.414. Run is found by multiplying Set by

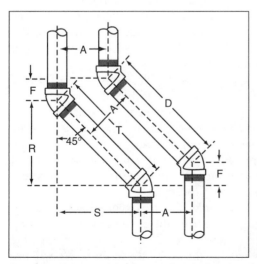

FIGURE 2.6 ■ Two-pipe 45° equal-spread offset.

1.000. The F measurement is found by multiplying the spread (A) by .4142. Remember that T and D are the same. Want to do the same exercise with a 60° setup? Why not?

To run a similar deal on 60° angles of equally- spaced offset pipes, you follow the same basic principles used in the previous example. Multiply the Set by 1.155 to find the Travel. Run is found by multiplying Set by .5773. The F measurement is found by multiplying the spread (A) by .5773. Remember that T and D are the same.

Need to find numbers for 22½° bends? Well, it's not difficult. To find figures for equally-spaced pipes with 22½° bends, multiply the Set by 2.613 to find the Travel. Run is found by multiplying Set by 2.414. The F measurement is found by multiplying the spread (A) by .1989. Remember that T and D are the same.

GETTING AROUND PROBLEMS

Getting around problems and obstacles is part of the plumbing and pipe fitting trades. Few jobs run without problems or obstacles. As any experienced piping contractor knows, there are always some obstructions in the preferred path of piping. Many times the obstacles are ductwork, but they can involve electrical work, beams, walls, and other objects are that not easily relocated. This means that the pipes must be rerouted. This section is going to deal with the mathematics required to compensate for immovable objects.

☑ *fast code* **fact**

Don't notch the bottom or top of floor or ceiling joists. Notches must not be closer than 1.5 inches of the top or bottom of a joist. When this is essential, the joist must be cut out and headed off.

Let me set the stage for a graphic example of getting around an overhead obstruction. Assume that you are bringing a pipe up and out of a concrete floor in a basement. There is a window directly above the pipe that you must offset around. The window was an afterthought. Having the pipe under the window was not a mistake in the groundworks rough-in. However, it is your job to move the pipe, without breaking up the floor, to get around the window.

In many cases, you might just cut the pipe off close to the floor, stick a 45° fitting on it, and bring a piece of pipe over to another 45° fitting. This is usually enough, but suppose you have a very tight space to work with and must make an extremely accurate measurement. Do you know how to do it? Imagine a situation where an engineer has indicated an exact location for the relocated riser. Can you hit the spot accurately? Do you know what type of formula to use in order to comply with the job requirements? If not, consider the following information as your ticket to success.

Our formula will involve three symbols. The first symbol will be an A, and it will be representative of the distance from the center of your 45° fitting to bottom edge of the window. The distance from the center of the rising pipe to the outside edge of the window will be known by the letter B. We will use the letter C to indicate the distance from the center of the Travel piece of

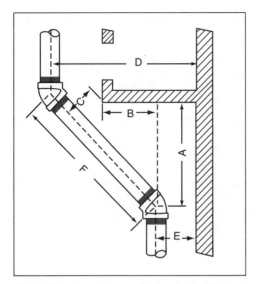

FIGURE 2.7 ▪ Pipe offsets around obstructions.

pipe from the edge of the window. Using E to indicate the distance of the center of the rising pipe from the right edge of the window and D to indicate the center of the offset rising pipe from the right edge of the window, we can . use the formula. Let's see how it works.

To find the distance from the bottom of the window to the starting point of the offset, you would take the distance from the center of the riser to the left edge of the window (B) and add the distance from the corner of the left window edge to the center of the pipe (C) times 1.414. The formula would look like this: $A = B + C \times 1.414$. Refer to Fig. 2.7 for an example. Now let's put measurement numbers into the formula.

Assume that you want to find A. Further assume that B is equal to one foot and that C is equal to six inches. The numerical formula would be like this: A = 12 inches (B) + 6 inches (C) × 1.414 = 12 inches + 8½ inches = 20½ inches. This would prove that the upper 45° fitting would be 20½ inches from the edge of the right edge of the window. As you can see, the actual procedure is not as difficult as the intimidation of using formulas might imply.

Round Obstacles

You've just seen how to get around what many would see as a typical problem. Most offsets are used to get around square or rectangular objects. But, what happens when you have to bypass a round object, such as a pressure tank? Don't worry, there is a simple way to get around most any problem, so let's talk about going around circular objects.

Okay, we have a pipe that has to rise vertically, but there is a horizontal expansion tank hanging in the ceiling that is blocking the path of our pipe. We have a very limited amount of space on either side of the tank to work within, so our measurements have to be precise. Assume that an eyeball measurement will not work in this case. So, let's set up the symbols that we will use in this formula.

Let's use the letter A to indicate the center of the offset rising pipe from the center of the expansion tank. The letter B will represent the center of the offset rising pipe from the edge of the tank. One-half of the diameter of the tank will be identified by the letter C. We will use the letter D to indicate the distance from the center line of the tank to the starting point of the offset. Additional information needed is that A = B + C and D = A × .4142. See Fig. 2.8 for a drawing to help you visualize the setup.

To put the letters into numbers, let's plug in some hypothetical numbers. Assign a number of 18 inches to C and eight inches to B. What is D? Here's how it works. A = B + C = 8 + 18 = 26 inches. D will equal A × .4142 = 26 × .4142 = 10¾ inches. This makes the center of the fitting 10¾ inches from the center of the tank.

ROLLING OFFSETS

Rolling offsets can be figured with a complex method or with a simplified method. Since I assume that you are interested in the most accurate information that you can get in the shortest amount of time, I will give you the simplified version. The results will be the same as the more complicated method, but you will not pull out as much hair or lose as much time as you would with the other exercise, and you will arrive at the same solution.

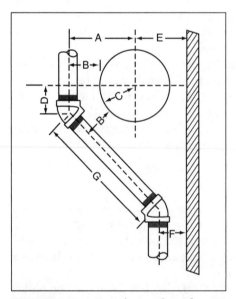

FIGURE 2.8 ■ **Starting point of a 45° offset around a tank.**

Angle	Constant
5⅝°	10.207
11¼°	5.125
22½°	2.613
30°	2.000
45°	1.414
60°	1.154

FIGURE 2.9 ■ Simplified method of figuring a rolling offset.

To figure rolling offsets simply, you will need a framing square, just a typical, steel, framing square. The corner of any flat surface is also needed, so that you can form a right angle. You will also need a simple ruler. The last tool needed is the table that I am providing in Fig. 2.9. This is going to be really easy, so don't run away. Let me explain how you will use these simple elements to figure rolling offsets.

Stand your framing square up on a flat surface. The long edge should be vertical and the short edge should be horizontal. The long, vertical section will be the Set, and the short, horizontal section will be the Roll. Your ruler will be used to tie the Set together with the Roll (Fig. 2.10 square and ruler). A constant will be needed to arrive at a solution, and you will find constants in the table I've provided in Fig. 2.9. Once again, the three main angles are addressed.

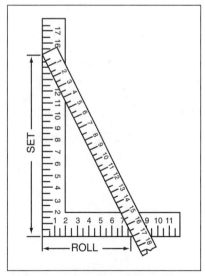

FIGURE 2.10 ■ Laying out a rolling offset with a steel square.

When you refer to Fig. 2.9, you will find that the constant for a 45° bend is 1.414. The number for a 60° bend is 1.154, and the constant for a 22½-bend is 2.613. If you were working with a 45° angle that had a Set of 15 inches and a Roll of eight inches, you would use your ruler to measure the distances between the two marks on the framing square. In this case, the measurement from the ruler would be 17 inches. You would multiply the 17-inch number by the 45° constant of 1.414 (found in Fig. 2.10) and arrive at a figure of 24¹/₃₂ of an inch. This would be the length of the pipe, from center to center, needed to make your rolling offset. Could it get any easier?

RUNNING THE NUMBERS

Running the numbers of pipe fitting is not always necessary to complete a job. If you have the experience and the eye to get the job done, without going through mathematical functions, that's great. I admit that I rarely have to use sophisticated math to figure out my piping layouts. But, I do know how to hit the mark right on the spot when I need to, and so should you. Accuracy can be critical. If you don't invest the time to learn the proper methods for figuring offsets, you may cut your career opportunities short. Believe me, you owe it to yourself to expand your knowledge. Sitting still can cost you. Reach out, as you are doing by reading this book, and expand your knowledge.

Some people see plumbers and pipe fitters as blue-collar workers. This may true. If it is, I'm proud to wear a blue collar. Yet, if you proceed in your career, you may own your own business, and this will, by society's standards, graduate you to a white collar. As far as I am concerned, the color of a person's collar has no bearing on the person's worth. Blue collar or white collar, individuals are what they are. We all bring something to the table. Yes, some people do prosper more than others, and education does play a role in most career advancements.

You may or may not need what you've learned in this chapter. However, knowing some simple math and having access to the tables in this chapter will probably give you an edge on many of the people you work with or compete with. Like it or not, making a living in today's world is competitive. So why not be as well prepared as possible? Okay, enough of the speech, let's move into the next chapter and study calculations that deal with welding fabrication and layout.

chapter 3

POTABLE WATER SYSTEMS CALCULATIONS

When you are sizing plumbing systems for potable water, you will work with fixture units. Ratings for fixture units are assigned to various plumbing fixtures by the plumbing codes. Not all of the ratings are the same from code to code. To perform an accurate sizing design, you must have factors for friction loss in pipe and because of valves, fittings, and water meters. A typical home is pretty easy to size, but when you are dealing with larger buildings and more plumbing, the procedure can become somewhat complicated.

If you were to design a potable water system for a home with two-bathrooms, you would probably run either a ³/₄-inch water service or a 1-inch water service into the building. Primary piping would be three-quarters of an inch in diameter, with branch piping having a diameter of one-half an inch. A rule of thumb is that not more than two fixtures should be served off of a single half-inch branch. This is a simple system without much of a load. But, what would happen if the building you were working with was an office building with four stories and a basement? There would be much more to consider, and I will provide you with a sizing example for this type of building in a few moments.

Many factors can come into play when sizing a water distribution system. The type of pipe or tubing being used is one factor. The friction loss among various

☞ been there **done that**

When you are dealing with large buildings, there will usually be detailed riser diagrams, blueprints, and specifications available to outline your work. House plans rarely show much more than fixture placement for plumbing. They frequently have wiring diagrams, but most don't show a piping schematic. Long story short, commercial buildings are usually laid out for the master plumber by a designer.

▶ *sensible* **shortcut**

Residential sizing is simple. Figure a 4-inch sewer pipe, plan on no more than two toilets on a 3-inch drain, and run three-quarter-inch water mains with no more than two water branches on a half-inch pipe. If you do this, you are unlikely to go wrong.

types of piping varies. Even the types of valves installed on the piping will make a difference in friction loss. Most important to sizing is the fixture-unit load on the system and the components of the system. In most cases, a job will start with one size of pipe and the pipes will grow increasingly smaller in diameter as they serve the various plumbing fixtures. Also, the rise of piping and the length of pipe runs will affect the sizing of a system.

Plumbing systems for most large jobs are designed by professionals who don't work as plumbers. However, when you are remodeling a building, adding onto an existing system, or working without detailed blueprints, the need for knowledge about pipe sizing may become very important. Sizing is also an element of most licensing exams for plumbers, so this is another good reason to learn and understand the principles used in sizing systems.

Two of the major plumbing codes have graciously agreed to allow me the use of excerpts from their codebooks to better show you rule and regulations pertaining to pipe sizing. Both of the codes offer sizing examples in their codebooks. Your local code may also offer similar sizing data. Once you have established numbers to work with, such as fixture-unit ratings, sizing a water system is a manageable task.

✓ *fast code* **fact**

When starting at a plumbing fixture, you can have the pipe size increase as you proceed downstream, but don't decrease the pipe size as you go downstream.

✓ *fast code* **fact**

Copper tubing is required to be supported at 6-foot intervals. This has been the case for years. If you are using PEX tubing, you will need to support the tubing at intervals that do not exceed 32 inches. Again, I want to stress that you must check your local, approved code for definitive answers.

SIZING WITH THE UNIFORM PLUMBING CODE

Sizing with data from the Uniform Plumbing Code is not too difficult. Allow me to give you some illustrations that are direct excerpts from the Uniform Plumbing Code. Look at the illustrations and try working through the sizing example that is provided (Fig. 3.1 through Fig. 3.17).

APPENDIX A

Because of the variable conditions encountered, it is impractical to lay down definite detailed rules of procedure for determining the sizes of water supply pipes in an appendix which must necessarily be limited in length. For a more adequate understanding of the problems involved, the reader is referred to Water-Distributing Systems for Buildings, Report BMS 79 of the National Bureau of Standards; and Plumbing Manual, Report BMS 66, also published by the National Bureau of Standards.

The following is a suggested order of procedure for sizing the water supply system.

A 1   Preliminary Information

A 1.1 Obtain the necessary information regarding the minimum daily service pressure in the area where the building is to be located.

A 1.2 If the building supply is to be metered, obtain information regarding friction loss relative to the rate of flow for meters in the range of sizes likely to be used. Friction-loss data can be obtained from most manufacturers of water meters. Friction losses for disk type meters may be obtained from Chart A-1.

CHART A-1
Friction Losses for Disk Type Water Meters

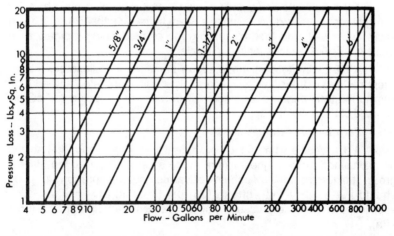

FIGURE 3.1 ■ Recommended rules for sizing the water supply system. (*Courtesy of The Uniform Plumbing Code*)

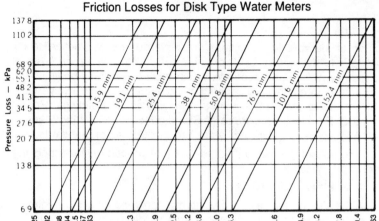

CHART A-1
(METRIC)
Friction Losses for Disk Type Water Meters

A 1.3 Obtain all available local information regarding the use of different kinds of pipe with respect both to durability and to decrease in capacity with length of service in the particular water supply.

A 2   Demand Load

A 2.1 Estimate the supply demand for the building main and the principal branches and risers of the system by totaling the fixture units on each, Table A-2, and then by reading the corresponding ordinate from Chart A-2 or A-3, whichever is applicable.

A 2.2 Estimate continuous supply demands in gallons per minute (liters per second) for lawn sprinklers, air conditioners, etc., and add the sum to the total demand for fixtures. The result is the estimated supply demand of the building supply.

A 3   Permissible Friction Loss

A 3.1 Decide what is the desirable minimum pressure that should be maintained at the highest fixture in the supply system. If the highest group of fixtures contains flushometer valves, the pressure for the group should not be less than fifteen (15) psi (103.4kPa). For flush tank supplies, the available pressure may not be less than eight (8) psi (55.1 kPa).

A 3.2 Determine the elevation of the highest fixture or group of fixtures above the water (street) main. Multiply this difference in elevation by forty-three hundredths (0.43). The result is the loss in static pressure in psi (pounds per square inch) (kPa).

FIGURE 3.2 ▪ Sizing rules. (*Courtesy of The Uniform Plumbing Code*)

A 3.3 Subtract the sum of loss in static pressure and the pressure to be maintained at the highest fixture from the average minimum daily service pressure. The result will be the pressure available for friction loss in the supply pipes, if no water meter is used. If a meter is to be installed, the friction loss in the meter for the estimated maximum demand should also be subtracted from the service pressure to determine the pressure loss available for friction loss in the supply pipes.

TABLE A-2
Demand Weight of Fixtures in Fixture Units [1]

Fixture Type[2]	Weight in Fixture Units[3] Private	Public	Minimum Connections Cold Water	Hot Water
Bathtub[4]	2	4	1/2	1/2
Bedpan washer		10	1	
Bidet	2	4	1/2	1/2
Combination sink and tray	3		1/2	1/2
Dental unit or cuspidor		1	3/8	
Dental lavatory	1	2	1/2	1/2
Drinking fountain	1	2	3/8	
Kitchen sink	2	4	1/2	1/2
Lavatory	1	2	3/8	3/8
Laundry tray (1or 2 compartments)	2	4	1/2	1/2
Shower, each head[4]	2	4	1/2	1/2
Sink, service	2	4	1/2	1/2
Urinal, pedestal		10	1	
Urinal (wall lip)		5	1/2	
Urinal stall		5	3/4	
Urinal with flush tank		3		
Wash sink, circular or multiple (each set of faucets)		2	1/2	1/2
Water closet:				
Flushometer-tank	3	5	3/8	
Flushometer valve	6	10	1	
Flush tank	3	5	3/8	

1 For supply outlets likely to impose continuous demands, estimate continuous supply separately and add to total demand for fixtures.

2 For fixtures not listed, weights may be assumed by comparing the fixture to a listed one using water in similar quantities and at similar rates.

3 The given weights are for total demand for fixtures with both hot and cold water supplies. The weights for maximum separate demands may be taken as seventy-five (75) percent of the listed demand for the supply.

4 Shower over bathtub does not add fixture unit to group.

FIGURE 3.3 ■ Fixture ratings. (*Courtesy of The Uniform Plumbing Code*)

TABLE A-3

Allowance in equivalent length of pipe for friction loss in valves and threaded fittings.*

Equivalent Length of Pipe for Various Fittings

Diameter of fitting (inches)	90° Standard Elbow	45° Standard Elbow	Standard Tee 90°	Coupling or Straight Run of Tee	Gate Valve	Globe Valve	Angle Valve
	Feet	Feet	Feet	Feet	Feet	Feet	Feet
3/8	1	0.6	1.5	0.3	0.2	8	4
1/2	2	1.2	3	0.6	0.4	15	8
3/4	2.5	1.5	4	0.8	0.5	20	12
1	3	1.8	5	0.9	0.6	25	15
1-1/4	4	2.4	6	1.2	0.8	35	18
1-1/2	5	3	7	1.5	1	45	22
2	7	4	10	2	1.3	55	28
2-1/2	8	5	12	2.5	1.6	65	34
3	10	6	15	3	2	80	40
4	14	8	21	4	2.7	125	55
5	17	10	25	5	3.3	140	70
6	20	12	30	6	4	165	80

TABLE A-3
(metric)

Equivalent Length of Pipe for Various Fittings

Diameter of fitting (inches)	90° Standard Elbow	45° Standard Elbow	Standard Tee 90°	Coupling or Straight Run of Tee	Gate Valve	Globe Valve	Angle Valve
	m	m	m	m	m	m	m
9.5	0.3	0.2	0.5	0.1	0.1	2.4	1.2
12.7	0.6	0.4	0.9	0.2	0.1	4.6	2.4
19.1	0.8	0.5	1.2	0.2	0.2	6.1	3.6
25.4	0.9	0.5	1.5	0.3	0.2	7.6	4.6
31.8	1.2	0.7	1.8	0.4	0.2	10.6	5.5
38.1	1.5	0.9	2.1	0.5	0.3	13.7	6.7
50.8	2.1	1.2	3	0.6	0.4	16.7	8.5
63.5	2.4	1.5	3.6	0.8	0.5	19.8	10.3
76.2	3	1.8	4.6	0.9	0.6	24.3	12.2
101.6	4.3	2.4	6.4	1.2	0.8	38	16.7
127	5.2	3	7.6	1.5	1	42.6	21.3
152.4	6.1	3.6	9.1	1.8	1.2	50.2	24.3

Allowances based on non-recessed threaded fittings. Use one-half (1/2) the allowances for recessed threaded fittings or streamline solder fittings.

FIGURE 3.4 ■ Friction loss factors. (*Courtesy of The Uniform Plumbing Code*)

A 3.4 Determine the developed length of pipe from the water (street) main to the highest fixture. If close estimates are desired, compute with the aid of Table A-3 the equivalent length of pipe for all fittings in the line from the water (street) main to the highest fixture and add the sum to the developed length. The pressure available for friction loss in pounds per square inch (kPa), divided by the developed lengths of pipe from the water (street) main to the highest fixture, times one hundred (100), will be the average permissible friction loss per one hundred (100) foot (30.4m) length of pipe.

A 4   Size of Building Supply

A 4.1 Knowing the permissible friction loss per one hundred (100 feet (30.4 m) of pipe and the total demand, the diameter of the building supply pipe may be obtained from Charts A-4, A-5, A-6, or A-7, whichever is applicable. The diameter of pipe on or next above the coordinate point corresponding to the estimated total demand and the permissible friction loss will be the size needed up to the first branch from the building supply pipe.

A 4.2 If copper tubing or brass pipe is to be used for the supply piping, and if the character of the water is such that only slight changes in the hydraulic characteristics may be expected, Chart A-4 may be used.

A 4.3 Chart A-5 should be used for ferrous pipe with only the most favorable water supply as regards corrosion and caking. If the water is hard or corrosive, Charts A-6 or A-7 will be applicable. For extremely hard water, it will be advisable to make additional allowances for the reduction of capacity of hot water lines in service.

A 5   Size of Principal Branches and Risers

A 5.1 The required size of branches and risers may be obtained in the same manner as the building supply by obtaining the demand load on each branch or riser and using the permissible friction loss computed in Section A 3.

A 5.2 Fixture branches to the building supply, if they are sized for the same permissible friction loss per one hundred (100) feet (30.4 m) of pipe as the branches and risers to the highest level in the building, may lead to inadequate water supply to the upper floor of a building. This may be controlled by: (1) selecting the sizes of pipe for the different branches so that the total friction loss in each lower branch is approximately equal to the total loss in the riser, including both friction loss and loss in static pressure; (2) throttling each such branch by means of a valve until the preceding balance is obtained; (3) increasing the size of the building supply and risers above the minimum required to meet the maximum permissible friction loss.

A 5.3 The size of branches and mains serving flushometer tanks shall be consistent with sizing procedures for flush tank water closets.

FIGURE 3.5 ▪ Sizing rules. (*Courtesy of The Uniform Plumbing Code*)

A 6   General

A 6.1 Velocities shall not exceed 10 ft/sec or the maximum values given in the appropriate Installation Standard, except as otherwise approved by the Administrative Authority.

A 6.2 If a pressure reducing valve is used in the building supply, the developed length of supply piping and the permissible friction loss should be computed from the building side of the valve.

A 6.3 The allowances in Table A-3 for fittings are based on nonrecessed threaded fittings. For recessed threaded fittings and streamlined soldered fittings, one-half (1/2) the allowances given in the table will be ample.

A 7   Example

A 7.1 Assume an office building of four (4) stories and basement; pressure on the building side of the pressure reducing valve of fifty-five (55) psi (379 kPa) (after an allowance for reduced pressure "fall off" at peak demand); an elevation of highest fixture above the pressure reducing valve of forty-five (45) feet (13.7 m); a developed length of pipe from the pressure reducing valve to the most distant fixture of two hundred (200) feet (60.8 m); and fixtures to be installed with flush valves for water closets and stall urinals as follows:

Example

Fixture Units and Estimated Demands

Kind of Fixtures	No. of Fixtures	Fixture units	Building Supply Demand gallons per minute (liters per second)	No. of Fixtures	Branch to hot-water system Fixture units	Demand gallons per minute (liters per second)
Water Closets	130	1,300				
Urinals	30	150				
Shower Heads	12	48		12	12 x 4 x	3/4 = 36 (2.3 L/s)
Lavatories	130	260		130	130 x 2 x	3/4 = 195 (12.3 L/s)
Service Sinks	27	108		27	27 x 4 x	3/4 = 81 5.1 L/s)
Total		1,866	313 (19.7 L/s)		312	106 (6.7 L/s)

Allowing for fifteen (15) psi (103.4 kPa) at the highest fixture under the maximum demand of three hundred and ten (310) gallons per minute (19.6 L/s), the pressure available for friction loss is found by the following:

FIGURE 3.6 ■ Fixture demands. (*Courtesy of The Uniform Plumbing Code*)

55 − [15 + (45 x 0.43)] = 20.65 psi
(metric) 379 − [103.4 + (13.7 x 9.8)] = 142.3 kPa

The allowable friction loss per one hundred (100) feet (30.4 m) of pipe is therefore:

100 x 20.65 ÷ 200 = 10.32 psi
(metric) 30.4 x 142.3 ÷ 60.8 = 71.1 kPa

If the pipe material and water supply are such that Chart A-5 applies, the required diameter of the building supply is three (3) inches (76.2 mm), and the required diameter of the branch to the hot water heater is two (2) inches (50.8 mm).

The sizes of the various branches and risers may be determined in the same manner as the size of the building supply or the branch to the hot water system by estimating the demand for the riser or branch from Charts A-2 or A-3, and applying the total demand estimate from the branch, riser or section thereof, to the appropriate flow chart.

FIGURE 3.7 ■ Sizing rules. (*Courtesy of The Uniform Plumbing Code*)

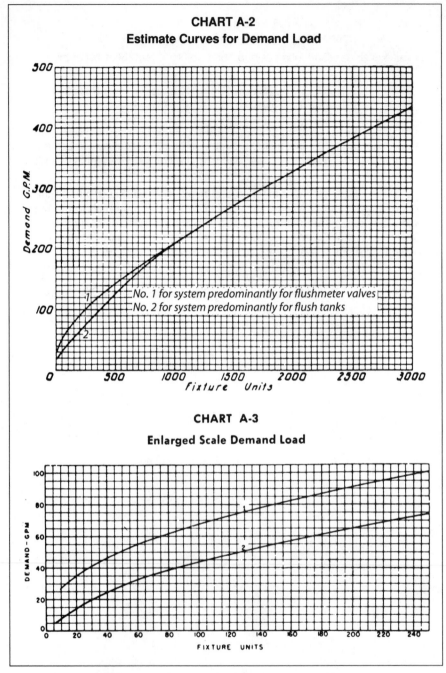

FIGURE 3.8 ■ Friction loss tables. (*Courtesy of The Uniform Plumbing Code*)

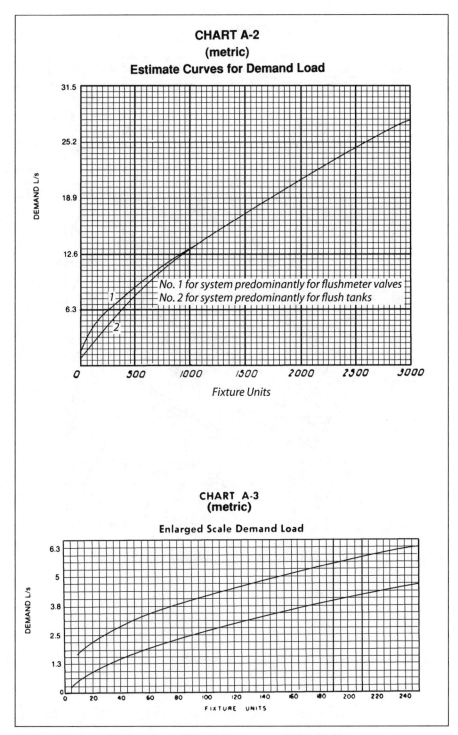

FIGURE 3.9 ■ Friction loss tables. (*Courtesy of The Uniform Plumbing Code*)

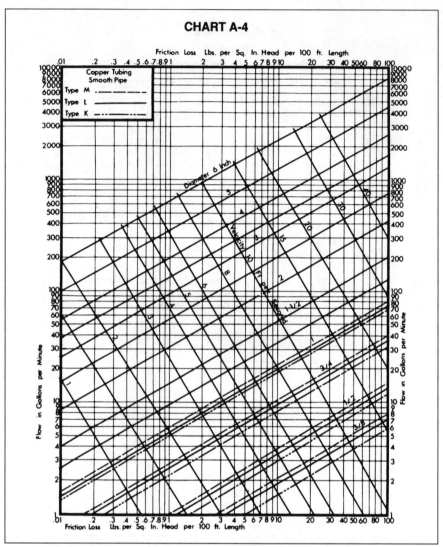

FIGURE 3.10 ■ Friction loss tables. (*Courtesy of The Uniform Plumbing Code*)

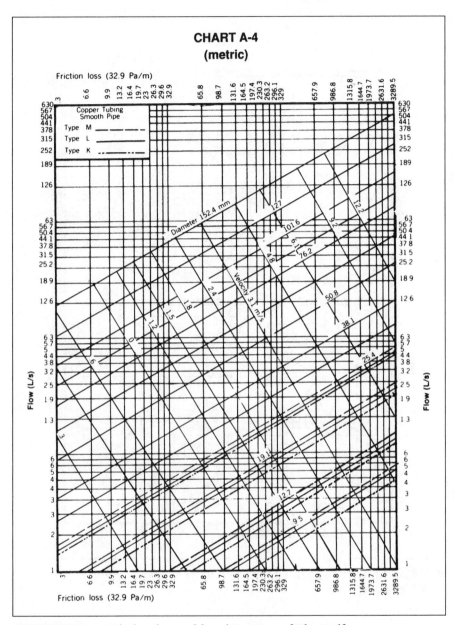

FIGURE 3.11 ■ Friction loss tables. (*Courtesy of The Uniform Plumbing Code*)

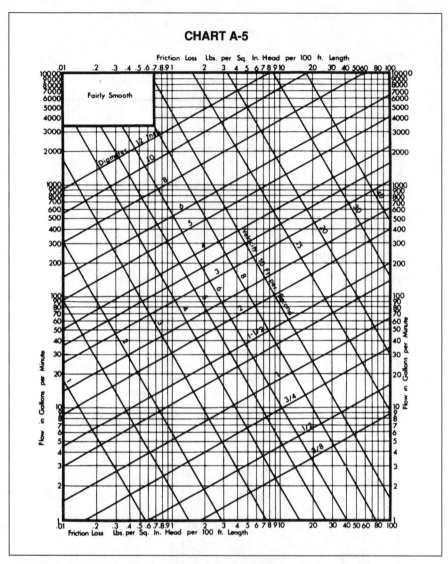

FIGURE 3.12 ■ Friction loss tables. (*Courtesy of The Uniform Plumbing Code*)

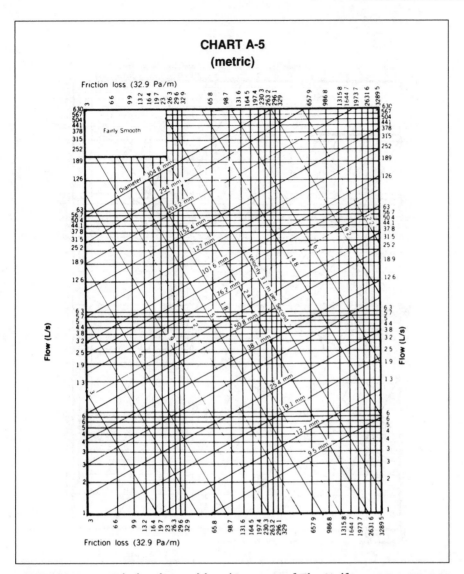

FIGURE 3.13 ▪ Friction loss tables. (*Courtesy of The Uniform Plumbing Code*)

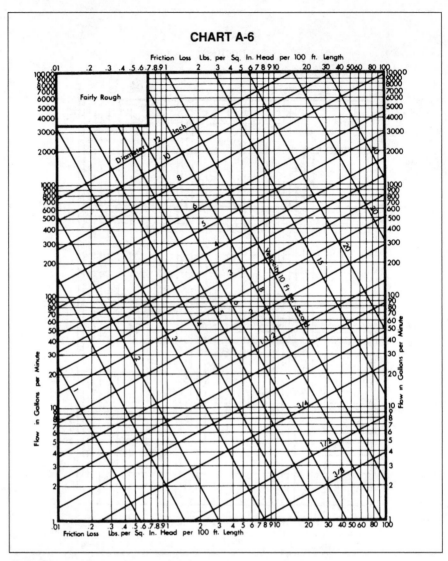

FIGURE 3.14 ■ Friction loss tables. (*Courtesy of The Uniform Plumbing Code*)

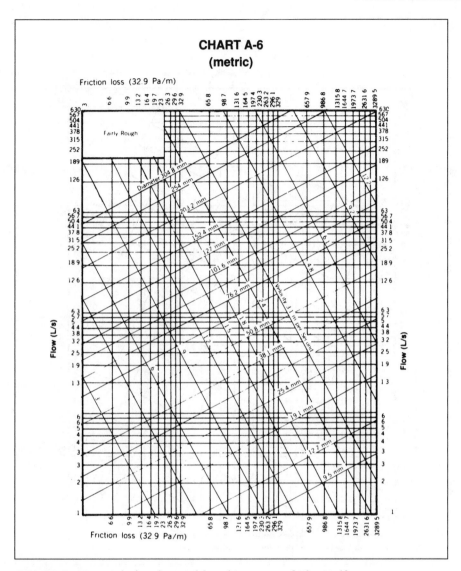

FIGURE 3.15 ▪ Friction loss tables. (*Courtesy of The Uniform Plumbing Code*)

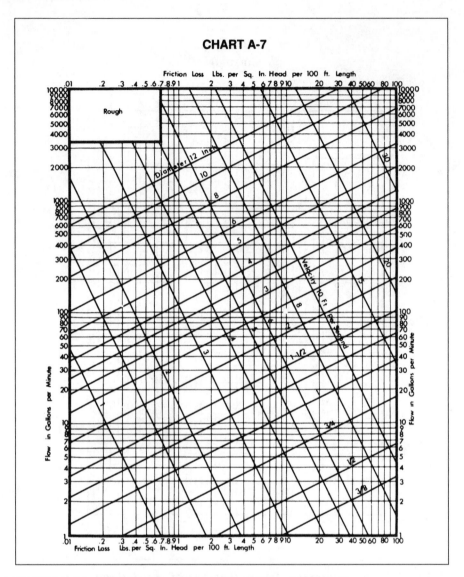

FIGURE 3.16 ■ Friction loss tables. (*Courtesy of The Uniform Plumbing Code*)

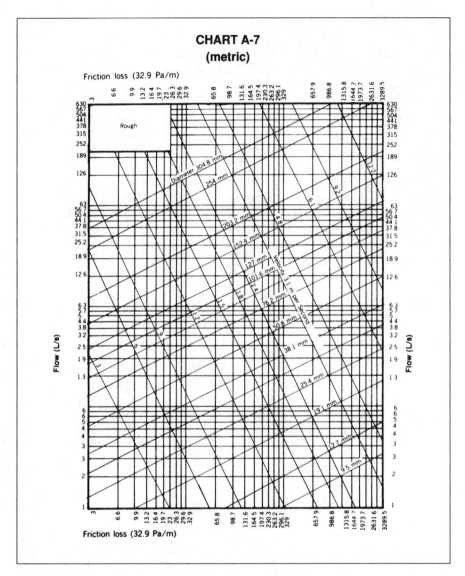

FIGURE 3.17 ■ Friction loss tables. (*Courtesy of The Uniform Plumbing Code*)

THE STANDARD PLUMBING CODE

The Standard Plumbing Code provides a sizing example in their codebook. The Standard Plumbing Code, the BOCA code, and the International Plumbing Code have joined together and your local code office may have any variation of these three codes in effect, so check your local codes carefully. A factory is the building chosen for their sizing exercise. The instructions provided in their sizing example and explanations is good. By using the tables provided and the formulas given, you can size the water distribution system for the building with minimal stress.

The responsibility of sizing a large water system may never be placed on your shoulders. Architects and engineers will probably design most of the systems that you install in large buildings. But, it does pay to understand the concepts behind sizing a system. Work through the examples I've provided above until you are comfortable with the procedure. Once you get the hang of it, sizing a system is not terribly difficult.

chapter 4

DRAIN-AND-SEWER CALCULATIONS

Doing calculations for drains and sewers is similar, in principle, to what you will use for vents in the next chapter. The process involves fixture units, the developed length of piping, sizing tables, and so forth. Paying attention to details is an important element in designing any type of system, and this certainly holds true when sizing drains and sewers. Moving too quickly and using the wrong sizing table can cause you a lot of trouble. When work is simplified, it sometimes seems so simple that it is taken too lightly. Don't make mistakes by not paying attention to footnotes and exclusions when you use sizing tables. If you read the tables carefully and apply them properly, sizing is not difficult.

Some plumbers get so accustomed to using sizing tables that they fail to think of code requirements that may make the tables inaccurate if all notes are not observed and followed. For example, let's say that you are sizing a sewer for a home. You might do your homework and find that the total number of fixture units is low enough that a 3-inch sewer can be used. This might be the case, but you could be setting yourself up for trouble. Accuracy in sizing pipes is essential to a job in more than one way. First, you have to draw riser diagrams and size the pipes for code approval. And, you need accurate sizing to price a job for bidding purposes.

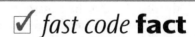

✓ *fast code* **fact**

Code requirements can be confusing. For example, you might find that a 3-inch pipe can carry a large number of fixture units. If you base your layout only on fixture units, you could make a big mistake. You are allowed a maximum number of toilets on a 3-inch pipe, even though the fixture-unit rating would be included. Read your code carefully and don't fall into the trap of doing your calculations quickly. Look at the "fine print" part of the code to stay out of trouble.

Let's say that you did believe that a home could get by with a 3-inch sewer. Assume that the home was a townhouse and that it was one of 300 in a project, where you were bidding the entire project. This means that you basically figure maybe four styles of houses and then apply the information to the entire project. The cost difference between 3-inch and 4-inch pipe doesn't seem like a lot when you are dealing with short runs. But, when you are dealing with 300 runs, even small differences in cost can add up quickly. So, what would happen if you figured a job for 3-inch sewers and wound up having to install 4-inch sewers? You or your employer would lose money, possibly a substantial sum.

▶ *sensible* **shortcut**

As a rule-of-thumb, don't plan on putting more than two toilets on a 3-inch drain. There are often conditions that will allow up to three toilets on a 3-inch drain, but I would suggest running a 4-inch pipe for three toilets if you want to keep it simple.

When we get further into this chapter you will see actual examples of how a mistake might be made when using sizing tables for drains and sewers. But, I'd like to point out a quick one now, so that you will keep your eyes open when we get to the sizing tables. Okay, it is very likely that the total fixture load for a 3-bathroom townhouse would be low enough to allow the use of a 3-inch sewer. If you were in a hurry, did a quick calculation of fixture units and scanned a sizing table, you might jump right at using 3-inch pipe to keep costs down. This would be a mistake. Why? Because even though a 3-inch pipe could handle the fixture units, most codes limit a 3-inch pipe to serving no more than two water closets in close proximity. If the townhouse has three toilets, a 4-inch sewer is likely to be needed.

As a teacher of plumbing courses, I've seen a number of experienced plumbers fall for this trap on some of the tests that I've created. The plumbers get into a rhythm and fail to think or to see the notes on the sizing charts and distance requirements. It's bad to miss a question on an exam, but it would be much worse to make the mistake in the real world of plumbing. By catching the plumbers in the classroom, I hope to make them aware of the crossover traps that can be embedded in the plumbing code. There are usually exceptions, options, and exclusions that can change the meaning of the code in certain situations. The 3-inch sewer is one excellent example of such pitfalls. You do have to pay attention to what you are doing when sizing systems.

TYPES OF SANITARY DRAINS

There are several types of sanitary drains. A building sewer is usually considered to be the main drain for a building that starts outside of the building and extends to a municipal sewer or private sewage-disposal system. Building drains are the primary drains inside of a building. Then there are branch intervals, horizontal branches, vertical stacks, and so forth. When you begin sizing a drainage system, you must make sure that you are using the proper sizing procedures for the type of drain or sewer that you are working with.

☞ been there **done that**

Learn plumbing terminology. Do you know the difference between a vent stack and a stack vent? If you don't, consult your codebook. Until you understand the language, you cannot do your job responsibly.

All types of drains and sewers can be calculated with a method that depends on the ratings of drainage fixture units. Fixture-unit ratings are established by local codes. A probability factor is built into the system. While a direct flow rate or discharge rate cannot be determined from the rating of fixture units, the fix-

▶ *sensible* **shortcut**

Use the tables in your codebook to calculate fixture units, load ratings, and pipe sizing. This is fast, easy, and accurate.

ture units are accurate enough to allow a sensible system to be designed in compliance with the plumbing code.

FIXTURE-UNIT TABLES

Fixture-unit tables are often used when sizing drains and sewers. The table in Figure 4.1 is an example of a table that expresses the maximum number of fixture units allowed on pipes of various sizes and with various amounts of fall. Before we go on, look at the category for 3-inch pipe, at a $\frac{1}{4}$-inch per foot fall. It says that you are allowed 27 drainage fixture units. But, notice the little number 2 next to the number of fixture units. That number indicates a note or exception. When you look at the bottom of the table, you will see that the note tells you that not more than two water closets can be carried on a 3-inch pipe. There are exceptions, but if you stick with this rule, you can't go wrong. This is one of the tables that I was telling you about earlier.

Diameter of pipe (in)	Fall in inches per foot			
	$\frac{1}{16}$	$\frac{1}{8}$	$\frac{1}{4}$	$\frac{1}{2}$
2			21	26
2½			24	31
3		20^2	27^2	36^2
4		180	216	250
5		390	480	575
6		700	840	1000
8	1400	1600	1920	2300
10	2500	2900	3500	4200
12	3900	4600	5600	6700
15	7000	8300	10,000	12,000

[1]Includes branches of the building drain. The minimum size of any building drain serving a water closet shall be 3".
[2]Not over two water closets.

FIGURE 4.1 ■ Allowable fixture-unit loads. (*Courtesy of McGraw-Hill*)

Fixture type	Fixture-unit value as load factors	Minimum size of trap (in)
Bathroom group consisting of water closet, lavatory, and bathtub or shower	6	
Bathtub (with or without overhead shower) or whirlpool attachments	2	1½
Bidet	2	Nominal 1½
Combination sink and tray	3	1½
Combination sink and tray with food disposal unit	4	Separate traps 1½
Dental unit or cuspidor	1	1¼
Dental lavatory	1	1¼
Drinking fountain	½	1
Dishwashing machine domestic	2	1½
Floor drains	1	2
Kitchen sink, domestic	2	1½
Kitchen sink, domestic with food waste grinder and/or dishwasher	3	1½
Lavatory	1	Small P.O. 1¼
Lavatory	2	Large P.O. 1½
Lavatory, barber, beauty parlor	2	1½
Lavatory, surgeon's	2	1½
Laundry tray (1 or 2 compartments)	2	1½
Shower stall, domestic	2	2
Showers (group) per head	3	
Sinks		
Surgeon's	3	1½
Flushing rim (with valve)	8	3
Service (trap standard)	3	3
Service (P trap)	2	2
Pot, scullery, etc.	4	1½
Urinal, pedestal, siphon jet, blowout	8	Note 6
Urinal, wall lip	4	Note 6
Urinal, washout	4	Note 6
Washing machines (commercial)		
Washing machine (residential)	3	2
Wash sink (circular or multiple) each set of faucets	2	Nominal 1½
Water closet, flushometer tank, public or private	3	Note 6
Water closet, private installation	4	Note 6
Water closet, public installation	6	Note 6

FIGURE 4.2 ■ Fixture-unit ratings. (*Courtesy of McGraw-Hill*)

The information in Figure 4.2 is representative of what you might find in your local codebook. This is the type of table that assigns specific ratings for fixture units on given fixtures. In cases where a known fixture is not listed, another type of table, like the one in Figure 4.3, is used to assign ratings for fixture units. Before we get too many tables in front of us, let's go over the three that you've just been introduced to.

The table in Figure 4.1 is easy enough to understand. If you find the size of the pipe you are working with, you can quickly ascertain the number of fixture units allowed on the pipe at a given grade. When you know the number of fixture units and the grade of the pipe, you can tell what size pipe is suitable. For example, a 4-inch sewer that is installed with a grade of one-quarter of an inch per foot can handle up to 216 fixture units, and that's a lot of drainage. Upgrading to a 6-inch pipe with the same grade will allow you to load the pipe with 840 drainage fixture units. That's all there is to that table.

The listings in Figure 4.2 are comprehensive and easy to understand. For example, a residential toilet is assigned a fixture-unit rating of four. A typical lavatory has a rating of one fixture unit. Domestic shower stalls are rated for two fixture units. If you add this up, you find that the three normal bathroom fixtures total a rating of seven fixture units. However, if you look at the top of the list, you will see that a bathroom group that consists of a toilet, lavatory, and bathtub or

> ▶ *sensible* **shortcut**
>
> When dealing with a residential property, a 4-inch sewer will handle all of the fixture units that could reasonably be installed in a home. Avoid installing a 3-inch sewer. Give homeowners the option of expansion by spending a little more money for the larger sewer.

shower has a rating of 6 fixture units. Wait a minute, that's one fixture unit less than the individual ratings for the same fixtures. What gives? In this case, assuming that all of the fixtures were being placed in the same bathroom, you could use the lower of the two ratings. Why? Because it is assumed that not all of the fixtures will be being used simultaneously if they are confined to a single room. The use of a table, like the one in Figure 4.2, makes sizing drains a lot easier.

There may be times when the fixture that you are seeking a rating for will not be listed on a fixture-unit table. If this is the case, you can use a table, like the one in Figure 4.3, to assign a rating for fixture units. For example, a fixture with a 2-inch drain that is not otherwise listed would be

> 🖘 been there **done that**
>
> As a young plumber, I believed that the code was the code. At the time, I worked in a metro area where there were numerous jurisdictions. Even being close together, I found out the hard way that not every city and county used the same code. Always check the local code in the area where you will be working to stay out of trouble.

Fixture drain or trap size (in)	Fixture-unit value
1¼ and smaller	1
1½	2
2	3
2½	4
3	5
4	6

FIGURE 4.3 ■ Allowable fixture units based on trap size. (*Courtesy of McGraw-Hill*)

| Diameter of pipe[5] (in) | Maximum no. of fixture units that may be connected to: | | | |
| | Any horizontal fixture branch[1,4] | One stack of 3 stories or 3 intervals maximum | More than 3 stories in height | |
			Total for stack	Total at one story or branch interval
1¼	1	2	2	1
1½	3	4	8	2
2	6	10	24	6
2½	12	20	42	9
3	20[2]	30[2]	60[3]	16[2]
4	160	240	500	90
5	360	540	1100	200
6	620	960	1900	350
8	1400	2200	3600	600
10	2500	3800	5600	1000
12	3900	6000	8400	1500
15	7000			

[1]Does not include branches of the building drain.
[2]Not over two water closets.
[3]Not over six water closets.
[4]50% less for battery vented fixture branches, no size reduction permitted for battery vented branches throughout the entire branch length.
[5]The minimum size of any branch or stack serving a water closet shall be 3".

FIGURE 4.4 ■ Maximum fixture units. (*Courtesy of McGraw-Hill*)

rated for three fixture units. A 4-inch drain would carry a rating of six fixture units. Pretty simple stuff, huh?

Some tables, like the one in Figure 4.4, deal with different piping arrangements. For example, the table in Figure 4.4 allows you to rate any horizontal branch stacks for multiple-story buildings and branch intervals. Notice that

Number of wet-vented fixtures	Diameter of vent stacks (in)
1 or 2 bathtubs or showers	2
3 to 5 bathtubs or showers	2½
6 to 9 bathtubs or showers	3
10 to 16 bathtubs or showers	4

FIGURE 4.5 ■ Vent sizing table. (*Courtesy of McGraw-Hill*)

Size of fixture drain (in)	Size of trap (in)	Fall (in/ft)	Max. distance from trap
1¼	1¼	¼	3 ft 6 in
1½	1¼	¼	5 ft
1½	1½	¼	5 ft
2	1½	¼	8 ft
2	2	¼	6 ft
3	3	⅛	10 ft
4	4	⅛	12 ft

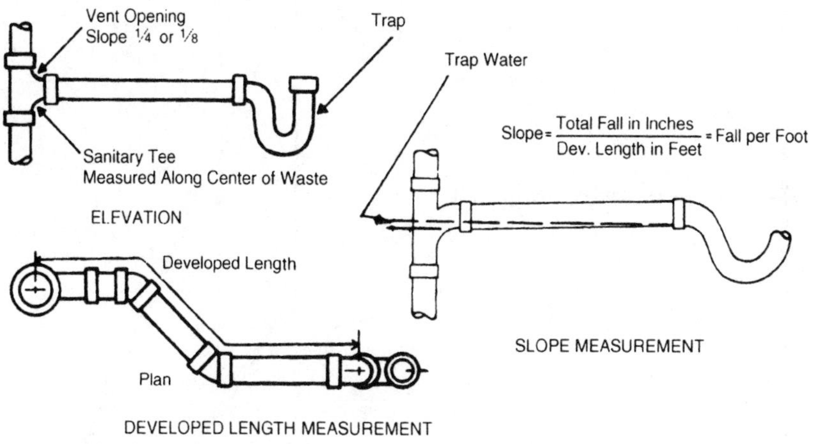

$$\text{Slope} = \frac{\text{Total Fall in Inches}}{\text{Dev. Length in Feet}} = \text{Fall per Foot}$$

FIGURE 4.6 ■ Trap-to-vent distances. (*Courtesy of McGraw-Hill*)

several of the ratings are marked with exclusions. This is the type of detailed information that you must be on the lookout for.

Suppose you are concerned about sizing a vent stack that will accommodate wet-vented fixtures? No problem, just use a table like the one in Figure 4.5. This table is so simple that it needs no explanation. Now, what if you need to know how long a trap arm may be? Refer to a table like the one in Figure 4.6 for the answers to your questions. Depending on trap size, the size of the fixture drain, and the amount of fall on the trap arm, you can choose a maximum length quickly.

Take a look at Figure 4.7. It is a riser diagram of a branch-interval detail. It is sometimes necessary to break a drainage system down into branch intervals for sizing. If you need to do this, you can refer to this drawing for a clear understanding of where branch intervals break and what they are. Figure 4.8 shows a stack with two branch intervals. To size a system like this, you must apply your sizing techniques to each individual branch and to the stack.

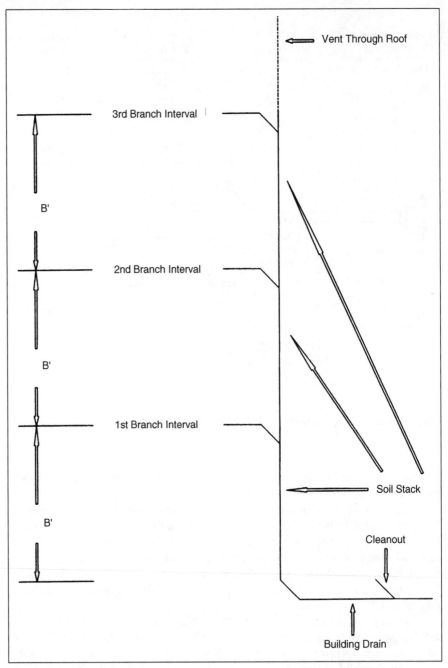

FIGURE 4.7 ■ Branch-interval detail. (*Courtesy of McGraw-Hill*)

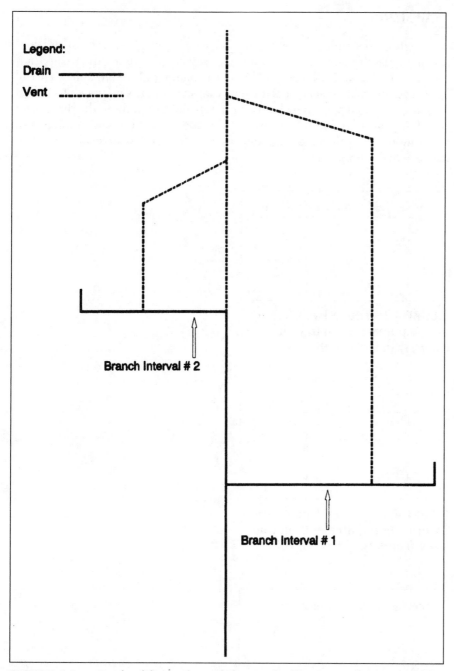

FIGURE 4.8 ■ Stack with two branch intervals. (*Courtesy of McGraw-Hill*)

TRAP SIZING

Trap sizing is a simple procedure. All you need is some basic information and a sizing table. If you know your trap size, you can determine the fixture-unit load that is allowable. When you know the number of fixture units that will be placed on a trap, you can decide on a trap size. There's not much to it. Figures 4.9, 4.10, and 4.11 show limits for fixture units on traps in the three main plumbing codes. If you notice, two of the codes have the same ratings, but one is more liberal than the other two. Remember to use your local code when doing actual sizing.

Trap size (in)	No. of fixture units
$1\frac{1}{4}$	1
$1\frac{1}{2}$	2
2	3
3	5
4	6

FIGURE 4.9 ■ Zone Two's fixture-unit requirements on trap sizes. (*Courtesy of McGraw-Hill*)

Trap size (in)	No. of fixture units
$1\frac{1}{4}$	1
$1\frac{1}{2}$	2
2	3
3	5
4	6

FIGURE 4.10 ■ Zone Three's fixture-unit requirements on trap sizes. (*Courtesy of McGraw-Hill*)

Trap size	No. of fixture units
$1\frac{1}{4}$	1
$1\frac{1}{2}$	3
2	4
3	6
4	8

FIGURE 4.11 ■ Zone One's fixture-unit requirements on trap sizes. (*Courtesy of McGraw-Hill*)

THE RIGHT PITCH

Having the right pitch on a pipe is necessary when complying with a plumbing code. The amount of pitch, or grade, on a pipe can affect its allowable length and fixture-unit load. You can use the tables in Figures 4.12, 4.13, and 4.14 as examples of how a local code might put rules in place for you to follow. The tables are easy to understand and use.

Pipe diameter (in)	Pitch (in/ft)
Under 3	¼
3–6	⅛
8 or larger	1/16

FIGURE 4.12 ■ Zone Three's minimum drainage-pipe pitch. (*Courtesy of McGraw-Hill*)

Pipe diameter (in)	Pitch (in/ft)
Under 3	¼
3 or larger	⅛

FIGURE 4.13 ■ Zone Two's minimum drainage-pipe pitch. (*Courtesy of McGraw-Hill*)

Pipe diameter (in)	Pitch (in/ft)
Under 4	¼
4 or larger	⅛

FIGURE 4.14 ■ Zone One's minimum drainage-pipe pitch. (*Courtesy of McGraw-Hill*)

✓ *fast code* **fact**

Be aware that S-traps are not legal for new installations and drum traps are usually, but not always, illegal. So are crown-vented traps. P-traps are the type most often used.

▶ *sensible* **shortcut**

Having too much grade on a pipe can be as bad as not having enough. If a drain pitches downward too hard, liquids will leave the pipe and suspend solids in the drain that could cause a stoppage. Maintain an even grade, usually one-quarter-of-an-inch per foot.

Pipe size (in)	Pipe grade (in/ft)	Maximum no. of fixture units
2	1/4	21
3	1/4	42*
4	1/4	216

*No more than two toilets may be installed on a 3-in building drain.

FIGURE 4.15 ■ Building-drain sizing table for Zone Three. (*Courtesy of McGraw-Hill*)

SIZING BUILDING DRAINS

Sizing building drains is simple when you have a sizing table and some basic information. Refer to Figure 4.15 for an example of a sizing table for a building drain. In this example, all pipes are based on a pitch of one-quarter of an inch per foot. A 3-inch pipe can carry up to 42 fixture units, but not more than two toilets. Tables like this one should be available in your local codebook.

A HORIZONTAL BRANCH

Let's talk about how you can size a horizontal branch. Bet you can guess that we are going to use a sizing table. Hey, they're easy, fast, and accurate, so why not use them? Look at Figure 4.16. This table shows you the maximum number of fixture units that may be placed on a single horizontal branch of a given size. If you look closely, you will see, once again, that not more than two toilets can be installed on a single 3-inch pipe that is installed horizontally. It should also be noted that the table does not represent the branches of a building drain and that other restrictions may apply if doing a series of battery venting.

Pipe size (in)	Maximum no. of fixture units on a horizontal branch
1 1/4	1
1 1/2	2
2	6
3	20†
4	160
6	620

*Table does not represent branches of the building drain, and other restrictions apply under battery-venting conditions.
†Not more than two toilets may be connected to a single 3-in horizontal branch. Any branch connecting with a toilet must have a minimum diameter of 3 in.

FIGURE 4.16 ■ Example of horizontal-branch sizing table in Zone Two. (*Courtesy of McGraw-Hill*)

Pipe size (in)	Fixture-unit discharge on stack from a branch	Total fixture units allowed on stack
1½	2	4
2	6	10
3	20*	48*
4	90	240

*No more than two toilets may be placed on a 3-in branch, and no more than six toilets may be connected to a 3-in stack.

FIGURE 4.17 ■ Stack-sizing table for Zone Three. (*Courtesy of McGraw-Hill*)

STACK SIZING

Stack sizing requires you to know the number of fixture units that will discharge into the stack from a single branch and the total number of fixture units that will be allowed on the stack. So, let's say that you have a stack with two branches. There is a bathroom group on each branch, and those two bathroom groups are all that will discharge into the stack. What size pipe is the smallest allowable for use as the stack? To figure this, use the table in Figure 4.17. So that you don't have refer back to the fixture-rate table, I will tell you that each bathroom group is rated for six fixture units. Well, we have two toilets, so we know the pipe size must be at least three inches in diameter. With 6 fixture units per branch we might get by with a 2-inch pipe if there were no toilets involved. But, toilets are involved and the total load on the stack will be 12 fixture units, so we have to go with a 3-inch pipe. For informational purposes, check out the sizing chart in Figure 4.18. Notice the difference in the number of fixture units allowed on a branch with Figure 4.18 when compared to Figure 4.17. There are two codes at work in these examples, and you can see that the difference for 4-inch pipe on a per-branch basis is 70 additional fixture units with one of the codes.

Pipe size (in)	Fixture-unit discharge on stack from a branch	Total fixture units allowed on stack
1½	3	4
2	6	10
3	20*	30*
4	160	240

*No more than two toilets may be placed on a 3-in branch, and no more than six toilets may be connected to a 3-in stack.

FIGURE 4.18 ■ Stack-sizing table for Zone Two. (*Courtesy of McGraw-Hill*)

Pipe size (in)	Fixture-unit discharge on stack from a branch	Total fixture units allowed on stack
1½	2	8
2	6	24
3	16*	60*
4	90	500

*No more than two toilets may be placed on a 3-in branch, and no more than six toilets may be connected to a 3-in stack.

FIGURE 4.19 ■ Stack-sizing tall stacks in Zone Two (stacks with more than three branch intervals). (*Courtesy of McGraw-Hill*)

SIZING TALL STACKS

Sizing tall stacks will require you to use different sizing tables. A tall stack is one that has more than three branch intervals. Figure 4.19 and Figure 4.20 will show you the basics needed to size tall stacks for two different codes. There are differences in the number of fixture units allowed between the two codes. Since the tables are so much like others we have used, I won't go into a lot of detail on them.

SUPPORTS

Supports for drainage systems are needed. The distance between supports varies with the type of pipe being used and the local code that you are working with. There are also differences between vertical and horizontal piping when you are designing your support placement. We could talk about this, but it would be faster and easier to just give you some reference tables to use when you need them. Figure 4.21 is for horizontal pipe with one code and

Pipe size (in)	Fixture-unit discharge on stack from a branch	Total fixture units allowed on stack
1½	2	8
2	6	24
3	20*	72*
4	90	500

*No more than two toilets may be placed on a 3-in branch, and no more than six toilets may be connected to a 3-in stack.

FIGURE 4.20 ■ Stack-sizing tall stacks in Zone Three (stacks with more than three branch intervals). (*Courtesy of McGraw-Hill*)

Support material	Maximum distance of supports (ft)
ABS	4
Cast iron	At each pipe joint*
Galvanized (1 in and larger)	12
Galvanized (¾ in and smaller)	10
PVC	4
Copper (2 in and larger)	10
Copper (1½ in and smaller)	6

*Cast-iron pipe must be supported at each joint, but supports may not be more than 10 ft apart.

FIGURE 4.21 ■ Horizontal pipe-support intervals in Zone One. (*Courtesy of McGraw-Hill*)

Figure 4.22 is for the same situation, but with a different code. Figure 4.23 deals with vertical pipes for one code, and Figure 4.24 shows vertical support requirements for a different code.

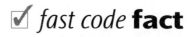

☑ *fast code* **fact**

Horizontal drainage pipe is required, normally, to be supported at a maximum interval of four feet.

Type of drainage pipe	Maximum distance of supports (ft)
ABS	4
Cast iron	At each pipe joint
Galvanized (1 in and larger)	12
PVC	4
Copper (2 in and larger)	10
Copper (1½ and smaller)	6

FIGURE 4.22 ■ Horizontal pipe-support intervals in Zone Two. (*Courtesy of McGraw-Hill*)

Type of drainage pipe	Maximum distance of supports*
Lead pipe	4 ft
Cast iron	At each story
Galvanized	At least every other story
Copper	At each story†
PVC	Not mentioned
ABS	Not mentioned

*All stacks must be supported at their bases.
†Support intervals may not exceed 10 ft.

FIGURE 4.23 ■ Vertical pipe-support intervals in Zone One. (*Courtesy of McGraw-Hill*)

Type of drainage pipe	Maximum distance of supports (ft)*
Lead pipe	4
Cast iron	At each story†
Galvanized	At each story‡
Copper (1¼ in and smaller)	4
Copper (1½ and larger)	At each story
PVC (1½ in and smaller)	4
PVC (2 in and larger)	At each story
ABS (1½ in and smaller)	4
ABS (2 in and larger)	At each story

*All stacks must be supported at their bases.
†Support intervals may not exceed 15 ft.
‡Support intervals may not exceed 30 ft.

FIGURE 4.24 ■ Vertical pipe-support intervals in Zone Two. (*Courtesy of McGraw-Hill*)

FITTINGS

As you are drawing your riser diagrams, you should keep in mind the fittings that will be used for changes in direction. There are three ways to change direction. Your pipe can go from horizontal to vertical, from vertical to horizontal, or from horizontal to horizontal. The fittings used in a drainage system to make these changes are regulated by the rules of the local plumbing code. As a rule-of-thumb, you can refer to Figure 4.25 for the common use and acceptance of fittings when changing directions. Again, always confirm local code requirements before committing to a job.

Type of fitting	Horizontal to vertical	Vertical to horizontal	Horizontal to horizontal
Sixteenth bend	yes	yes	yes
Eighth bend	yes	yes	yes
Sixth bend	yes	yes	yes
Quarter bend	yes	no	no
Short sweep	yes	yes	no
Long sweep	yes	yes	yes
Sanitary tee	yes	no	no
Wye	yes	yes	yes
Combination wye and eighth bend	yes	yes	yes

FIGURE 4.25 ■ Allowable fittings to accommodate changes in direction. (*Courtesy of McGraw-Hill*)

RISER DRAWINGS

Riser drawings are used when figuring out drainage systems, just as they are used with vent systems. I want to give you some sample riser diagrams to look over. The drawings will show you what your drawings might look like. Drains are drawn with solid lines, while vents are indicated by broken lines.

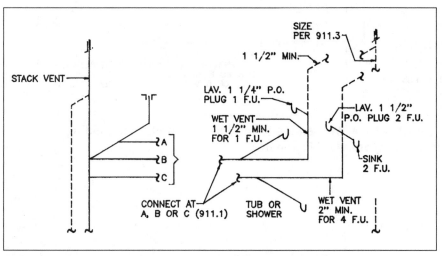

FIGURE 4.26 ■ Wet venting top floor single bath group. (*Courtesy of Standard Plumbing Code*)

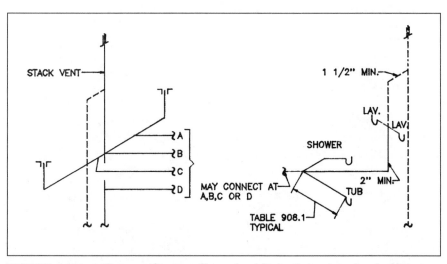

FIGURE 4.27 ■ Wet venting top floor double bath back to back. (*Courtesy of Standard Plumbing Code*)

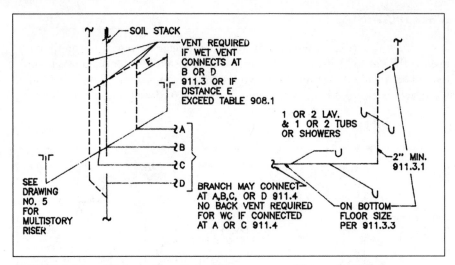

FIGURE 4.28 ■ Wet venting lower floors on multistory buildings. (*Courtesy of Standard Plumbing Code*)

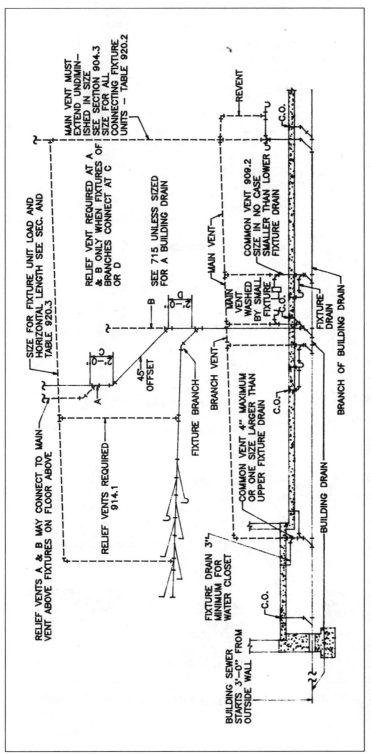

FIGURE 4.29 ■ Riser diagram. *(Courtesy of Standard Plumbing Code)*

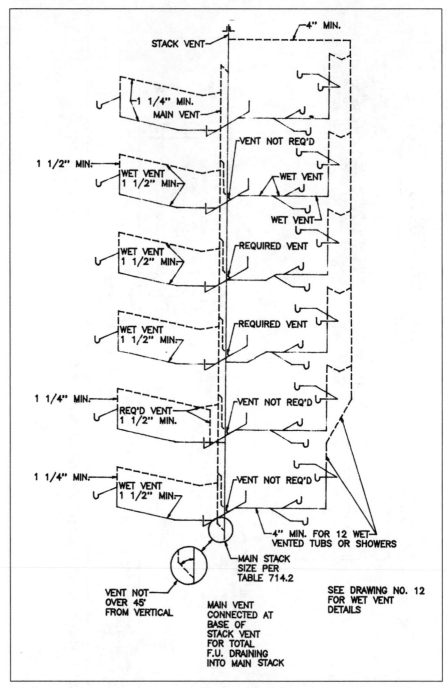

FIGURE 4.30 ■ Multistory wet venting. (*Courtesy of Standard Plumbing Code*)

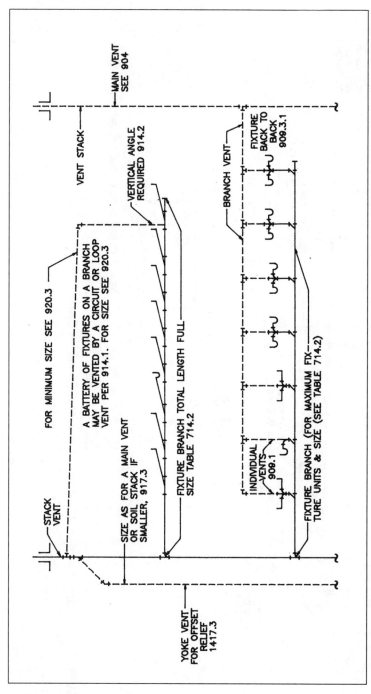

FIGURE 4.31 ■ Riser diagram. *(Courtesy of Standard Plumbing Code)*

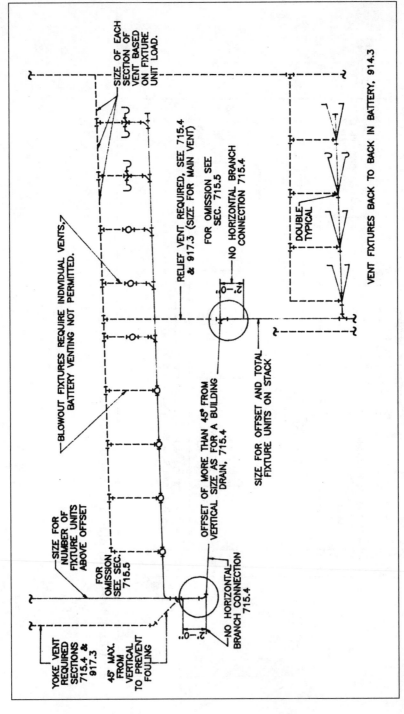

FIGURE 4.32 ■ Riser diagram. *(Courtesy of Standard Plumbing Code)*

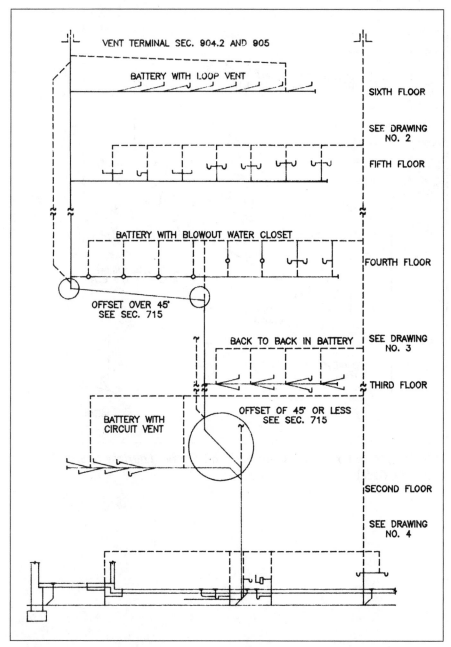

FIGURE 4.33 ■ Drainage waste and vent reference diagram. (*Courtesy of Standard Plumbing Code*)

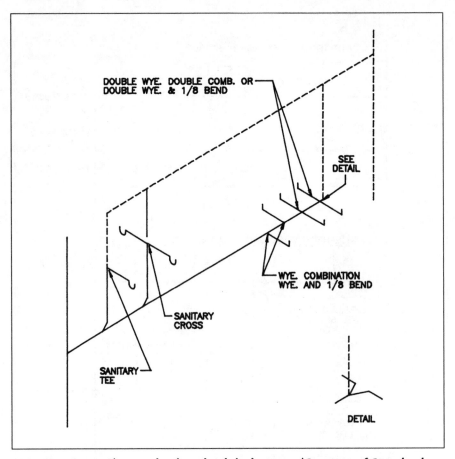

FIGURE 4.34 ■ Fixtures back-to-back in battery. (*Courtesy of Standard Plumbing Code*)

chapter 5

VENT SYSTEM CALCULATIONS

The calculations of vent systems are not very difficult to understand. There are sizing tables that you can use to compute pipe sizes. Plumbers need to understand the types of vents and master plumbers must be able to assign pipe sizes to them. The task is important, but not really very difficult for experienced plumbers. In many cases, engineers and architects are the ones who design plumbing systems. This is okay. But, it is not always the case. Having the ability to size a vent system is something that is basically a requirement for a master plumber's license. Of course, there are many types of plumbing vents. We can talk about dry vents, wet vents, branch vents, yoke vents, and lots of other types of vents. Before we get into the sizing of vents, I want to identify typical types of plumbing vents. This may be old news to you. If it is, skip past the section and jump right into the sizing information. But, if you are not versed in the full arrangement of vents, you might enjoy the illustrations that I will provide to indicate the basic ingredients of various types of vents.

 fast code **fact**

When it comes to code compliance, you must apply the same serious consideration to vents that you would to a drainage system. Hangers must be in their proper placement, the grade must be set correctly, and the vents must be tested for leaks.

TYPES OF VENTS

Types of vents are numerous. Do you know what an island vent is? How much do you know about relief vents? Depending upon your level of knowledge in the plumbing trade, you might be aware of all types of vents, but not all readers are. Before we jump into sizing examples, I'd like to make sure that all of my readers are aware of the various types of vents. With this in mind, I will provide a number of vent drawings for readers to devour. So, let's get on with the visual examples of various vent types.

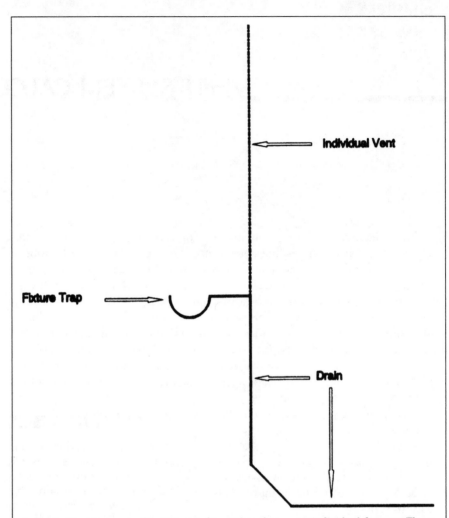

Individual vents are, as the name implies, vents that serve individual fixtures. These vents only vent one fixture, but they may connect into another vent that will extend to the open air. Individual vents do not have to extend from the fixture being served to the outside air, without joining another part of the venting system, but they must vent to open air space.

Sizing an individual vent is easy. The vent must be at least one-half the size of the drain it serves, but it may not have a diameter of less than 1¼ in. For example, a vent for a 3-in drain could, in most cases, have a diameter of 1½ in. A vent for a 1½-in drain may not have a diameter of less than 1¼ in.

FIGURE 5.1 ■ Individual vents. *(Courtesy of McGraw-Hill)*

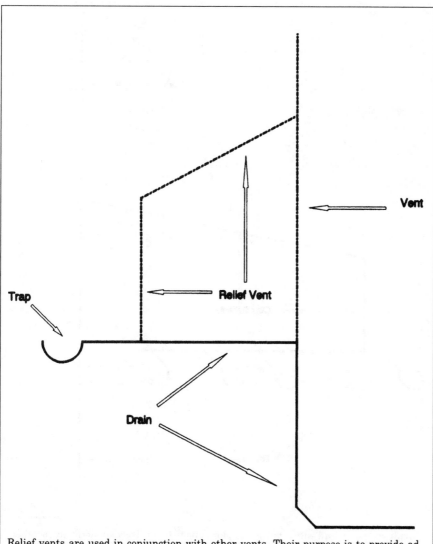

Relief vents are used in conjunction with other vents. Their purpose is to provide additional air to the drainage system when the primary vent is too far from the fixture. Relief vents must be at least one-half the size of the pipe it is venting. For example, if a relief vent is venting a 3-in pipe, the relief vent must have a 1½-in or larger diameter.

FIGURE 5.2 ■ Relief vents. (*Courtesy of McGraw-Hill*)

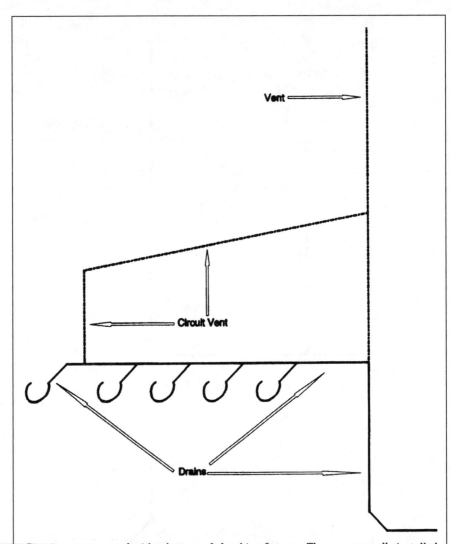

Circuit vents are used with a battery of plumbing fixtures. They are normally installed just before the last fixture of the battery. Then, the circuit vent is extended upward to the open air or tied into another vent that extends to the outside. Circuit vents may tie into stack vents or vent stacks. When sizing a circuit vent, you must account for its developed length. But in any event, the diameter of a circuit vent must be at least one-half the size of the drain it is serving.

FIGURE 5.3 ■ Circuit vents. (*Courtesy of McGraw-Hill*)

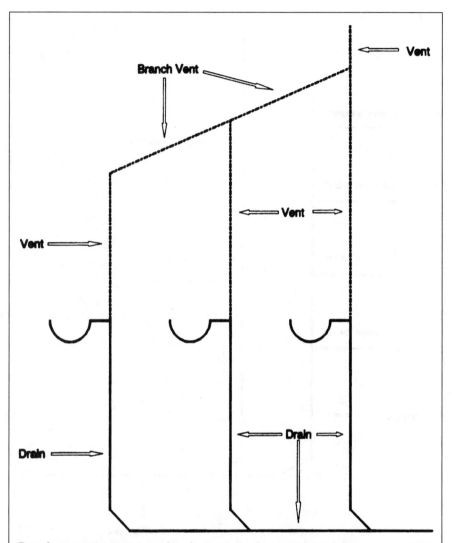

Branch vents are vents extending horizontally that connect multiple vents together. Branch vents are sized with the developed-length method, just as you were shown in the examples above. A branch vent or individual vent that is the same size as the drain it serves is unlimited in the developed length it may obtain. Be advised, zone two and zone three use different tables and ratings for sizing various types of vents; zone one uses the same rating and table for all normal venting situations.

FIGURE 5.4 ▪ Branch vents. (*Courtesy of McGraw-Hill*)

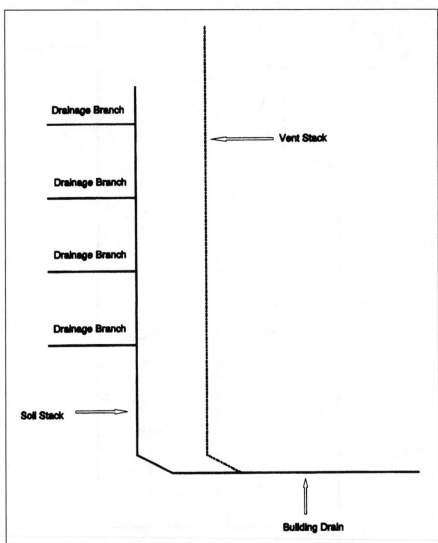

A vent stack is a pipe used only for the purpose of venting. Vent stacks extend upward from the drainage piping to the open air, outside of a building. Vent stacks are used as connection points for other vents, such as branch vents. A vent stack is a primary vent that accepts the connection of other vents and vents an entire system.

FIGURE 5.5 ■ Vent stack. (*Courtesy of McGraw-Hill*)

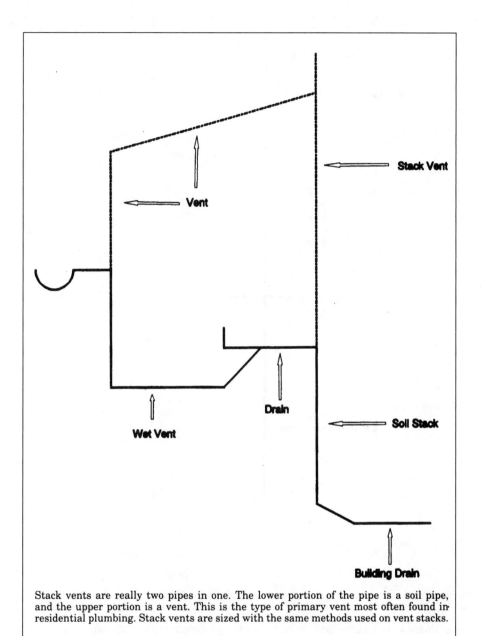

Stack vents are really two pipes in one. The lower portion of the pipe is a soil pipe, and the upper portion is a vent. This is the type of primary vent most often found in residential plumbing. Stack vents are sized with the same methods used on vent stacks.

FIGURE 5.6 ■ Stack vents. (*Courtesy of McGraw-Hill*)

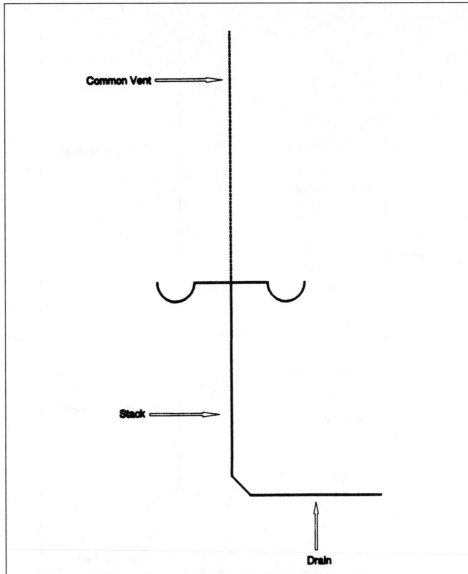

Common vents are single vents that vent multiple traps. Common vents are only allowed when the fixtures being served by the single vent are on the same floor level. Zone one requires the drainage of fixtures being vented with a common vent to enter the drainage system at the same level. Normally, not more than two traps can share a common vent, but there is an exception in zone three. Zone three allows you to vent the traps of up to three lavatories with a single common vent. Common vents are sized with the same techniques applied to individual vents.

FIGURE 5.7 ▪ Common vents. (*Courtesy of McGraw-Hill*)

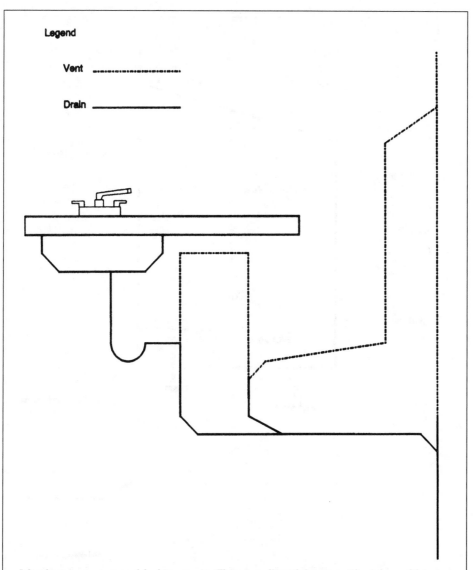

Legend

Vent ·······························

Drain _____

Island vents are unusual looking vents. They are allowed for use with sinks and lavatories. The primary use for these vents is with the trap of a kitchen sink, when the sink is placed in an island cabinet. Sometimes pictures speak louder than words.

As you can see from the figure, island venting may take a little getting used to. Notice that the vent must rise as high as possible under the cabinet before it takes a U-turn and heads back downward. Since this piping does not rise above the flood-level rim of the fixture, it must be considered a drain. Fittings approved for drainage must be used in making an island vent.

FIGURE 5.8 ▪ Island vents. *(Courtesy of McGraw-Hill)*

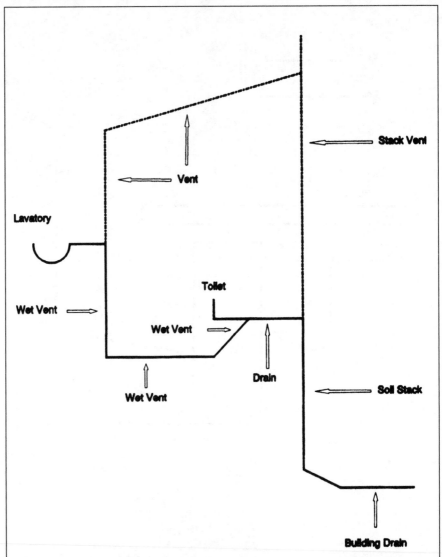

Wet vents are pipes that serve as a vent for one fixture and a drain for another. Wet vents, once you know how to use them, can save you a lot of money and time. By effectively using wet vents you can reduce the amount of pipe, fittings, and labor required to vent a bathroom group or two.

FIGURE 5.9 ■ Wet vents. (*Courtesy of McGraw-Hill*)

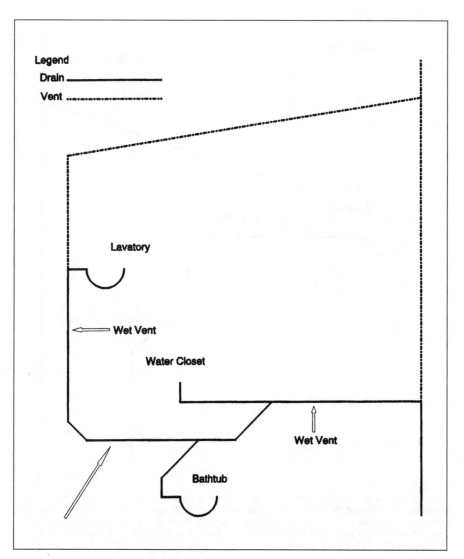

FIGURE 5.10 ■ Web venting a bathroom group. (*Courtesy of McGraw-Hill*)

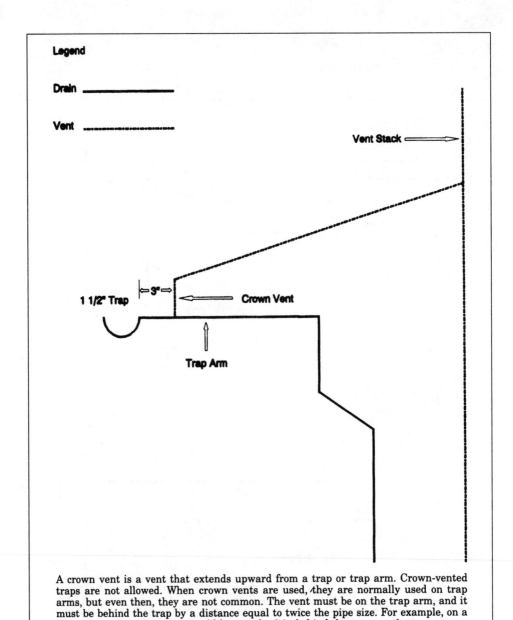

Legend

Drain _____

Vent ·····················

Vent Stack ⟵⟶

1 1/2" Trap   ⟻3"⟼   ⟸ Crown Vent

Trap Arm

A crown vent is a vent that extends upward from a trap or trap arm. Crown-vented traps are not allowed. When crown vents are used, they are normally used on trap arms, but even then, they are not common. The vent must be on the trap arm, and it must be behind the trap by a distance equal to twice the pipe size. For example, on a 1½-in trap, the crown vent would have to be 3 in behind the trap, on the trap arm.

FIGURE 5.11 ■ Crown vents. (*Courtesy of McGraw-Hill*)

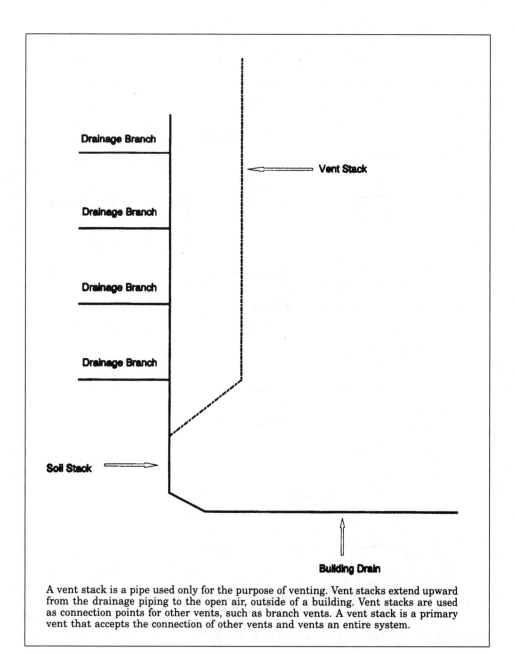

A vent stack is a pipe used only for the purpose of venting. Vent stacks extend upward from the drainage piping to the open air, outside of a building. Vent stacks are used as connection points for other vents, such as branch vents. A vent stack is a primary vent that accepts the connection of other vents and vents an entire system.

FIGURE 5.12 ■ Vent stacks. (*Courtesy of McGraw-Hill*)

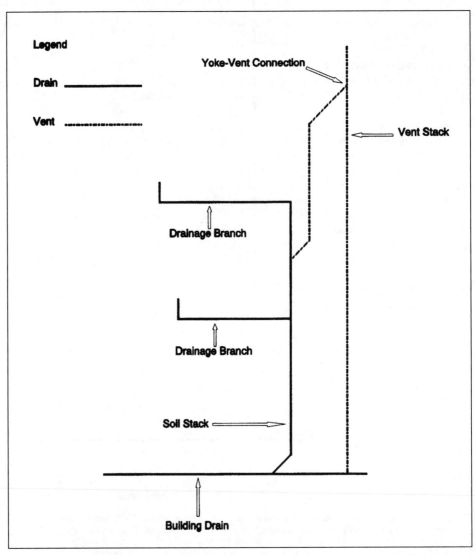

FIGURE 5.13 ■ Yoke vent. (*Courtesy of McGraw-Hill*)

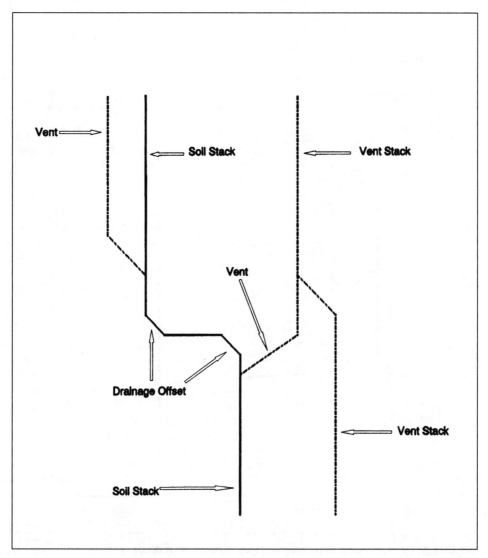

FIGURE 5.14 ■ Example of venting drainage offsets. (*Courtesy of McGraw-Hill*)

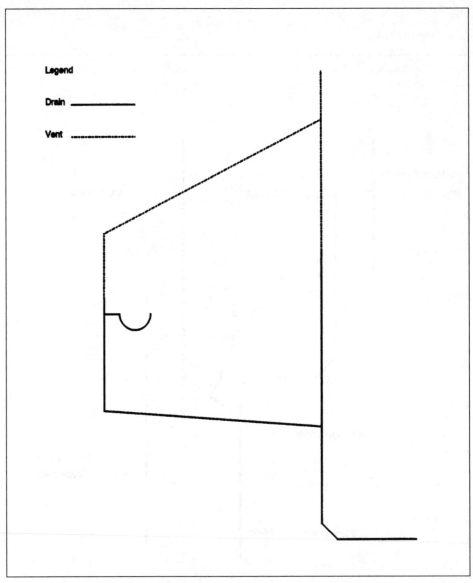

Legend

Drain ———————

Vent ·····················

FIGURE 5.15 ■ Graded-vent connection. (*Courtesy of McGraw-Hill*)

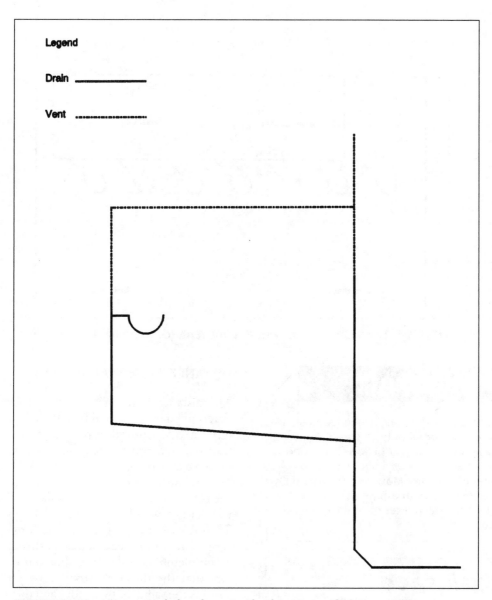

FIGURE 5.16 ■ Zone one's level-vent rule. (*Courtesy of McGraw-Hill*)

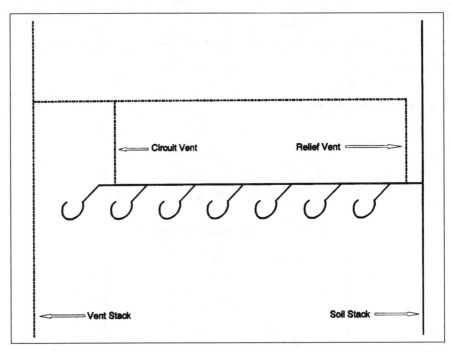

FIGURE 5.17 ■ Circuit vent with a relief vent. (*Courtesy of McGraw-Hill*)

▶ sensible **shortcut**

When in doubt about if a vent is needed, install one. If you are not sure about the size of a vent, make it larger than what you believe is needed. For example, a toilet requires a vent with a minimum diameter of 2 inches. Most other residential fixtures can be vented with a pipe that has a diameter of 1.5 inches. All homes must have at least one 3-inch vent.

☑ *fast code* **fact**

When computing the distance from a trap to a vent, you must use the developed length of the entire piping. For example, you would measure from the trap along the length of the drainpipe to the point where the vent is connected to the drain. In other words, you can't measure on a short angle from a vent to a trap; you must measure the total length of the pipe used as a drain.

DISTANCE FROM TRAP TO VENT

The distance from a trap to a vent is determined by local plumbing code requirements. Allowable distances are usually given either in text or in tables within a codebook. There can be a significant difference from one code to the other. To illustrate this, I'd like you to refer to Figures 5.18, 5.19, and 5.20. The tables you see in these illustrations represent differences between three major plumbing codes. You should notice that the distances from traps to vents is the same for two codes and different for one code. You must refer to the plumbing code that is being enforced in your area for specific sizing requirements. The information provided here is representative of the types of charts, tables, and information you will likely work with, but it is not necessarily the code that you will be

Grade on drain pipe (in)	Size of trap arm (in)	Maximum distance between trap and vent (ft)
¼	1¼	2½
¼	1½	3½
¼	2	5
¼	3	6
¼	4 and larger	10

FIGURE 5.18 ■ Trap-to-vent distances in Zone One. (*Courtesy of McGraw-Hill*)

Grade on drain pipe (in)	Fixture's drain size (in)	Trap size (in)	Maximum distance between trap and vent (ft)
¼	1¼	1¼	3½
¼	1½	1¼	5
¼	1½	1½	5
¼	2	1½	8
¼	2	2	6
⅛	3	3	10
⅛	4	4	12

FIGURE 5.19 ■ Trap-to-vent distances in Zone Two. (*Courtesy of McGraw-Hill*)

Grade on drain pipe (in)	Fixture's drain size (in)	Trap size (in)	Maximum distance between trap and vent (ft)
¼	1¼	1¼	3½
¼	1½	1¼	5
¼	1½	1½	5
¼	2	1½	8
¼	2	2	6
⅛	3	3	10
⅛	4	4	12

FIGURE 5.20 ■ Trap-to-vent distances in Zone Three. (*Courtesy of McGraw-Hill*)

working with. Our intent here is to learn how to size systems, so consider the information here as a learning tool, rather than a code ruling.

SIZING TABLES

Sizing tables are often used when sizing vent pipes (Fig. 5.21). There can be many different types of tables to use during a sizing procedure. For example,

Drain pipe size (in)	Drain pipe size (in/ft)	Vent pipe size (in)	Maximum developed length of vent pipe (ft)
1½	¼	1¼	Unlimited
1½	¼	1½	Unlimited
2	¼	1¼	290
2	¼	1½	Unlimited
3	¼	1½	97
3	¼	2	420
3	¼	3	Unlimited
4	¼	2	98
4	¼	3	Unlimited
4	¼	4	Unlimited

FIGURE 5.21 ■ Vent sizing table for Zone Three (for use with individual, branch, and circuit vents for horizontal drain pipes). (*Courtesy of McGraw-Hill*)

you might use one table to size a vent stack (Fig. 5.22) and another table to size a wet stack vent (Fig. 5.23). Some codes might use one table for both types of vents (Fig. 5.24). Then you might have a different table to use when sizing branch vents or circuit vents (Fig. 5.25). Battery vents may require a different table (Fig. 5.26). Once you have a sizing table to work with, sizing a vent system is not a complicated process.

Wet-vented fixtures	Stack size required (inches)
1 to 2 Bathtubs or showers	2
3 to 5 Bathtubs or showers	2½
6 to 9 Bathtubs or showers	3
10 to 16 Bathtubs or showers	4

FIGURE 5.22 ■ Sizing a vent stack for wet-venting in Zone Two. (*Courtesy of McGraw-Hill*)

Pipe size of stack (inches)	Fixture-unit load on stack	Maximum length of stack
2	4	30
3	24	50
4	50	100
6	100	300

FIGURE 5.23 ■ Sizing a wet stack vent in Zone Two. (*Courtesy of McGraw-Hill*)

Drain pipe size (inches)	Fixture-unit load on drain pipe	Vent pipe size (inches)	Maximum developed length of vent pipe (feet)
1½	8	1¼	50
1½	8	1½	150
1½	10	1¼	30
1½	10	1½	100
2	12	1½	75
2	12	2	200
2	20	1½	50
2	20	2	150
3	10	1½	42
3	10	2	150
3	10	3	1040
3	21	1½	32
3	21	2	110
3	21	3	810
3	102	1½	25
3	102	2	86
3	102	3	620
4	43	2	35
4	43	3	250
4	43	4	980
4	540	2	21
4	540	3	150
4	540	4	580

FIGURE 5.24 ■ Vent sizing for Zone Three (for use with vent stacks and stack vents). (*Courtesy of McGraw-Hill*)

Drain pipe size (inches)	Drain pipe grade per foot (inches)	Vent pipe size (inches)	Maximum developed length of vent pipe (feet)
1½	¼	1¼	Unlimited
1½	¼	1½	Unlimited
2	¼	1¼	290
2	¼	1½	Unlimited
3	¼	1½	97
3	¼	2	420
3	¼	3	Unlimited
4	¼	2	98
4	¼	3	Unlimited
4	¼	4	Unlimited

FIGURE 5.25 ■ Vent sizing for Zone Three (for use with individual, branch, and circuit vents for horizontal drain pipes. (*Courtesy of McGraw-Hill*)

Soil or waste pipe diam. (in)	Maximum no. fixture units	Diameter of circuit or loop vent (in)					
		1½	2	2½	3	4	5
2	3	15	40				
2½	6	10	30				
3	10	—	20	40	100		
4	80	—	7	20	52	200	
5	180	—	—	—	16	70	200

1 in = 25.4 mm
1 ft = 0.3048 m

FIGURE 5.26 ■ Battery vent sizing table (maximum horizontal length (ft)). (*Courtesy of Standard Plumbing Code*)

☞ been there done that

Don't allow yourself to become confused when sizing vents. Pay attention to the tables that you are using and make sure that you are working with the right table for the type of vent you are sizing.

A SIZING EXERCISE

Let's do a sizing exercise to illustrate how the tables from a codebook might be used to determine the size of piping needed for various vents. When you set up a vent system, you must know how far a vent is allowed to be from the trap it is serving. If you look at Figure 5.27, you will see the requirements for one of the major plumbing codes. The table is easy enough to understand. If you have a fixture drain that has a diameter of 1.5 inches and a trap size of 1.5 inches, with a grade of a quarter of an inch per foot, the trap may be as much as five feet from the vent. With this particular code, the distance would remain the same, even if the trap size was only one and a quarter inches in diameter, so long as the drain remains as a 1.5 inch diameter.

Size of fixture drain (in)	Size of trap (in)	Fall (in/ft)	Max. distance from trap
1¼	1¼	¼	3 ft 6 in
1½	1¼	¼	5 ft
1½	1½	¼	5 ft
2	1½	¼	8 ft
2	2	¼	6 ft
3	3	⅛	10 ft
4	4	⅛	12 ft

1 in = 25.4 mm
1 ft = 0.3048 m

FIGURE 5.27 ■ Distance of fixture trap from vent. (*Courtesy of Standard Plumbing Code*)

If the size of the fixture drain was three inches in diameter, with a 3-inch trap, and one-eighth of an inch of fall per foot, the vent could be up to 10 feet from the trap. Obviously, this type of table is easy to understand and to work with.

Vent sizing is based on developed length. This is the measured distance of all pipe used in the system. Measurements are taken on a center-to-center basis. You can see in Figure 5.28 how the measurements are assessed. Once you know the developed length of a vent, you can use a sizing chart to determine the minimum diameter of the vent pipe. The sizing of a vent or vent system is not difficult. Let me show you how it's done.

Look at Figure 5.29. This is a chart designed for sizing individual and branch vents serving horizontal soil and waste branches. As you look at the table, you will see two types of abbreviations. The abbreviation shown as NP means "Not Permitted". When you see the abbreviation of UL, it means "Unlimited". Aside from these two clarifications, the table pretty much speaks for itself. Try to find the answer to the question I'm about to give you. Assume that you have a drain that has a 2-inch diameter. The amount of fall on the pipe is set at one quarter of an inch per foot. You want to run a vent with a diameter of 1.5 inches. How far can you run the vent in that size? The answer is that there is no limit to the length of the vent run. But, suppose you wanted the vent diameter to be 1.25 inch, how far could it go? A vent of this size

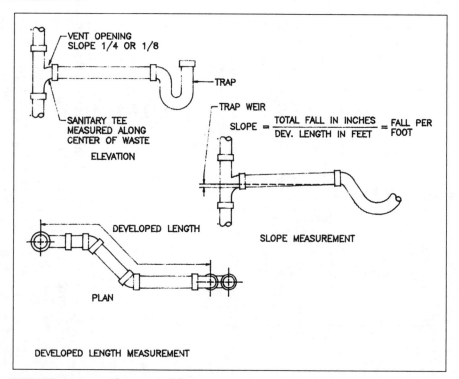

FIGURE 5.28 ■ Distance of fixture trap from vent. (*Courtesy of Standard Plumbing Code*)

Diameter of horizontal drainage piping (inches)[1]	Slope of horizontal drainage branch (inches per foot)	Maximum developed length (feet): measured from the connection to the most remote fixture served to either the outside termination or the connection to either the stack vent or vent stack diameter of vent (in)									
		1¼	1½	2	2½	3	4	5	6	8	10
1¼	¼	UL[3]									
	½	UL									
1½	¼	UL	UL								
	½	UL	UL								
2	¼	874	UL	UL							
	½	437	UL	UL							
2½	¼	286	756	UL	UL						
	½	143	378	UL	UL						
3	⅛	NP	606	UL	UL	UL					
	¼	NP	303	UL	UL	UL					
	½	NP	152	704	UL	UL					
4	⅛	NP	NP	666	UL	UL	UL				
	¼	NP	NP	333	UL	UL	UL				
	½	NP	NP	166	548	UL	UL				
5	⅛	NP	NP	218	716	UL	UL	UL			
	¼	NP	NP	109	358	948	UL	UL			
	½	NP	NP	54	179	474	UL	UL			

FIGURE 5.29 ■ Individual and branch vent sizing table for horizontal soil and waste branches. (*Courtesy of Standard Plumbing Code*)

could run for a developed length of 874 feet, which is far more than you would be likely to run it.

Now assume that you have a 3-inch drain and you want to run a vent that has a diameter of 1.25 inches. How far can it go? It can't be used. This is indicated by the NP symbol. The reason for this is that the vent must be at least half the size, in diameter, of what the drain being served is. There are exceptions to this rule as the size of drains become larger. This means that the smallest vent diameter allowed for a 3-inch drain is a 1.5-inch vent. A vent of this size could run for 606 feet. When you deal with large-diameter drains, you have to move up to larger vent sizes to achieve unlimited runs of distance. This can be seen in Figure 5.30. You would use the table in Figure 5.30 in the same way that we used the previous sizing table.

STACK VENTS, VENT STACKS, AND RELIEF VENTS

Stack vents, vent stacks, and relief vents require more information for sizing. Specifically, you must know the fixture-unit load on a drain before you can determine vent sizing and developed lengths. Once you have calculated the fixture units, a sizing table can be used to give you your sizing information. The table shown in Figure 5.31 is the type of table that would be used to define the requirements of stack vents, vent stacks, and relief vents. Like the other tables, this one is self-explanatory. To prove this, size the diameter and maximum length of a vent stack that will serve 20 fixture units with a drain diameter of two inches. Assume that you want your vent pipe to have a diameter of one and a half inches. The answer is that your vent size is okay, as long as you don't extend it more than 50 feet. If you need more distance, increase the size of the vent to a 2-inch diameter and feel free to run the vent up to 150 feet. The table is easy to use, but you must be able to calculate the load of fixture units. How will you do that? I'll show you.

Most codebooks will provide you with some form or a chart or table that identifies fixture-units for drainage piping. A table like the one in Figure 5.32 is quite helpful. By looking at such a table, you can quickly determine the load, in terms of fixture units, that an individual fixture puts on a drain. For example, a bidet carries a fixture-unit rating of 2. A drinking fountain is rated for one half of a fixture unit. A residential water closet is worth 4 fixture units. By using this type of table to assign fixture-unit ratings to all fixtures being served by a drain, you can then arrive at a number to use with the vent-sizing table. For fixtures that are not listed, you can use a generic table, like the one in Figure 5.33, to assign ratings for fixture units.

✓ *fast code* **fact**

The terminal height of a vent that penetrates a roof can change from region to region. For example, one state may require a vent to extend 12 inches above a roof while another state may require the vent to extend 24 inches above a roof. This type of difference can be due to snow loads on roofs in some areas.

Diameter of horizontal drainage piping (inches)[1]	Slope of horizontal drainage branch (inches per foot)	Maximum developed length (feet): measured from the connection to the most remote fixture served to either the outside termination or the connection to either the stack vent or vent stack diameter of vent (in)									
		1¼	1½	2	2½	3	4	5	6	8	10
6	⅛	NP	NP	87	287	780	UL	UL	UL	—	—
	¼	NP	NP	44	144	380	UL	UL	UL	—	—
	½	NP	NP	22	72	190	883	UL	UL	—	—
8	⅛	NP	NP	21	68	180	835	UL	UL	UL	
	¼	NP	NP	10	34	90	417	UL	UL	UL	
	½	NP	NP	NP	17	45	209	687	UL	UL	
10	⅛	NP	NP	NP	22	59	273	898	UL	UL	UL
	¼	NP	NP	NP	11	29	136	449	UL	UL	UL
	½	NP	NP	NP	NP	15	68	224	594	UL	UL
12	⅛	NP	NP	NP	NP	24	108	360	953	UL	UL
	¼	NP	NP	NP	NP	12	55	180	476	UL	UL
	½	NP	NP	NP	NP	NP	27	90	238	UL	UL
15	⅛	NP	NP	NP	NP	NP	36	118	311	UL	UL
	¼	NP	NP	NP	NP	NP	18	59	156	723	UL
	½	NP	NP	NP	NP	NP	NP	29	78	362	UL

1 foot = 304.8 mm
1 inch per foot = 83.3 mm/m

UL means unlimited length = Actual values in excess of 1,000 feet.
NP means not permitted.

Notes:
1. Size of fixture drain served for sizing individual vents and size of horizontal branch served for sizing branch vents that serve more than one individual vent.

FIGURE 5.30 ■ Individual and branch vent sizing table for horizontal soil and waste branches. (*Courtesy of Standard Plumbing Code*)

Size of soil or waste stack (in)	Fixture units connected	Diameter of vent required (in)								
		1¼	1½	2	2½	3	4	5	6	8
1¼	2	30								
1½	8	50	150							
1½	10	30	100							
2	12	30	75	200						
2	20	26	50	150						
2½	42	—	30	100	300					
3	10	—	30	100	200	600				
3	30	—	—	60	200	500				
3	60	—	—	50	80	400				
4	100	—	—	35	100	260	1000			
4	200	—	—	30	90	250	900			
4	500	—	—	20	70	180	700			
5	200	—	—	—	35	80	350	1000		
5	500	—	—	—	30	70	300	900		
5	1100	—	—	—	20	50	200	700		
6	350	—	—	—	25	50	200	400	1300	
6	620	—	—	—	15	30	125	300	1100	
6	960	—	—	—	—	24	100	250	1000	
6	1900	—	—	—	—	20	70	200	700	
8	600	—	—	—	—	—	50	150	500	1300
8	1400	—	—	—	—	—	40	100	400	1200
8	2200	—	—	—	—	—	30	80	350	1100
8	3600	—	—	—	—	—	25	60	250	800
10	1000	—	—	—	—	—	—	75	125	1000
10	2500	—	—	—	—	—	—	50	100	500
10	3800	—	—	—	—	—	—	30	80	350
10	5600	—	—	—	—	—	—	25	60	250

1 in = 25.4 mm
1 ft = 0.3048 m

FIGURE 5.31 ■ Maximum length of stack vents, vent stacks, and relief vents. (*Courtesy of Standard Plumbing Code*)

Fixture type	Fixture-unit value as load factors	Minimum size of trap (in)
Bathroom group consisting of water closet, lavatory, and bathtub or shower	6	
Bathtub[1] (with or without overhead shower) or whirlpool attachments	2	1½
Bidet	2	Nominal 1½
Combination sink and tray	3	1½
Combination sink and tray with food disposal unit	4	Separate traps 1½
Dental unit or cuspidor	1	1¼
Dental lavatory	1	1¼
Drinking fountain	½	1
Dishwashing machine[2] domestic.	2	1½
Floor drains[5]	1	2
Kitchen sink, domestic	2	1½
Kitchen sink, domestic with food waste grinder and/or dishwasher	3	1½
Lavatory[4]	1	Small P.O. 1¼
Lavatory[4]	2	Large P.O. 1½
Lavatory, barber, beauty parlor	2	1½
Lavatory, surgeon's	2	1½

FIGURE 5.32 ■ Fixture units per fixture or group. (*Courtesy of Standard Plumbing Code*)

WET VENTING

Wet venting is popular, but a little different when it comes to sizing the vents. Tables can still be used for this type of sizing. Look at Figure 5.34 for an example of a table that might be used to size a wet stack vent. Another type of table that you might encounter is shown in Figure 5.35. This table is intended for use in sizing a vent stack for wet venting.

Keep in mind that not all plumbing codes are the same, and they may present their information differently. It is also important to remember that requirements may be different.

Fixture type	Fixture-unit value as load factors	Minimum size of trap (in)
Laundry tray (1 or 2 compartments)	2	1½
Shower stall, domestic	2	2
Showers (group) per head[2]	3	
Sinks		
Surgeon's	3	1½
Flushing rim (with valve)	8	3
Service (trap standard)	3	3
Service ("p" trap)	2	2
Pot, scullery, etc.[2]	4	1½
Urinal, pedestal, siphon jet, blowout	8	Note 6
Urinal, wall lip	4	Note 6
Urinal, Washout	4	Note 6
Washing machines (commercial)[3]		
Washing machine (residential)	3	2
Wash sink (circular or multiple) each set of faucets	2	Nominal 1½
Water closet, flushometer tank, public or private	3	Note 6
Water closet, private installation	4	Note 6
Water closet, public installation	6	Note 6

1 in = 25.4 mm

Notes:

1. A showerhead over a bathtub or whirlpool bathtub attachments does not increase the fixture value.

2. See Figures 12.33 and 12.34 for methods of computing unit value of fixtures not listed in Figure 12.32 or for rating of devices with intermittent flows.

3. See Figure 12.33.

4. Lavatories with 1¼ or 1½-inch trap have the same load value; larger P.O. plugs have greater flow rate.

5. Size of floor drain shall be determined by the area of the floor to be drained. The drainage fixture unit value need not be greater than 1 unless the drain receives indirect discharge from plumbing fixtures, air conditioner or refrigeration equipment.

6. Trap size shall be consistent with fixture type as defined in industry standards.

FIGURE 5.32 ▪ (*Continued*) Fixture units per fixture or group. (*Courtesy of Standard Plumbing Code*)

Fixture drain or trap size (in)	Fixture-unit value
1¼ and smaller	1
1½	2
2	3
2½	4
3	5
4	6
1 in = 25.4 mm	

FIGURE 5.33 ■ **Fixtures not listed.** (*Courtesy of Standard Plumbing Code*)

Stack pipe size	Fixture-unit load on stack	Maximum length of stack (ft)
2	4	30
3	24	50
4	50	100
6	100	300

FIGURE 5.34 ■ **Table for sizing a wet stack vent in Zone Two.** (*Courtesy of McGraw-Hill*)

No. of fixtures	Vent-stack size requirements (in)
1–2 bathtubs or showers	2
3–5 bathtubs or showers	2½
6–9 bathtubs or showers	3
10–16 bathtubs or showers	4

FIGURE 5.35 ■ **Table for sizing a vent stack for wet venting in Zone Two.** (*Courtesy of McGraw-Hill*)

SUMP VENTS

Sump vents, the ones used to vent a sump system, are calculated on a basis of a pump's discharge capacity. Tables are often provided for this type of sizing. See Figure 5.37 for an example of such a table. Using a table like this one, you can quickly and easily size a vent for a sump. As long as you know the discharge rate of the pump being used in the sump, the rest of the work is simple.

Drain pipe size (in)	Fixture-unit load on drain pipe	Vent pipe size (in)	Maximum developed length of vent pipe (ft)
1½	8	1¼	50
1½	8	1½	150
1½	10	1¼	30
1½	10	1½	100
2	12	1½	75
2	12	2	200
2	20	1½	50
2	20	2	150
3	10	1½	42
3	10	2	150
3	10	3	1040
3	21	1½	32
3	21	2	110
3	21	3	810
3	102	1½	25
3	102	2	86
3	102	3	620
4	43	2	35
4	43	3	250
4	43	4	980
4	540	2	21
4	540	3	150
4	540	4	580

FIGURE 5.36 ■ Vent sizing table for Zone Three. (*Courtesy of McGraw-Hill*)

Discharge rate of pump (gallons per minute)	Maximum developed length of vent (in feet)	Diameter of vent (in inches)
10	NL	1½
10	NL	2
10	NL	3
10	NL	4
20	NL	1½
20	NL	2
20	NL	3
20	NL	4
40	160	1½
40	NL	2
40	NL	3
40	NL	4
60	75	1½
60	270	2
60	NL	3
60	NL	4
80	41	1½
80	150	2
80	NL	3
80	NL	4

FIGURE 5.37 ■ Sizing sump pumps.

✓ *fast code* **fact**

If a vent is run up the outside wall of a building and is exposed to weather, the pipe must be protected from freezing. One way of doing this that is generally accepted is to enlarge the diameter of the vent to prevent condensation from freezing and ultimately sealing the vent pipe with ice.

SUPPORTING A VENT SYSTEM

Supporting a vent system is another element of a system design. The spacing allowed for support varies from code to code and with the type of pipe being used in the vent system. The tables below will show you some examples of recommended minimums for support spacing.

Support material	Maximum distance between supports (ft)
ABS	4
Cast iron	At each pipe joint*
Galvanized	12
Copper (1½ in and smaller)	6
PVC	4
Copper (2 in and larger)	10

*Cast-iron pipe must be supported at each joint, but supports may not be more than 10 ft apart.

FIGURE 5.38 ■ Horizontal pipe-support intervals in Zone Two. (*Courtesy of McGraw-Hill*)

Type of vent pipe	Maximum distance between supports
Lead pipe	4 ft
Cast iron	At each story
Galvanized	At least every other story
Copper	At each story†
PVC	Not mentioned
ABS	Not mentioned

*All stacks must be supported at their bases.
†Support intervals may not exceed 10 ft.

FIGURE 5.39 ■ Vertical pipe-support intervals in Zone One. (*Courtesy of McGraw-Hill*)

Type of vent pipe	Maximum distance between supports (ft)
ABS	4
Cast iron	At each pipe joint
Galvanized	12
PVC	4
Copper (2 in and larger)	10
Copper (1½ in and smaller)	6

FIGURE 5.40 ■ Horizontal pipe-support intervals in Zone Two. (*Courtesy of McGraw-Hill*)

Type of vent pipe	Maximum distance between supports (ft)
Lead pipe	4
Cast iron	At each story†
Galvanized	At each story‡
Copper (1¼ in)	4
Copper (1½ in and larger)	At each story
PVC (1½ in and smaller)	4
PVC (2 in and larger)	At each story
ABS (1½ in and smaller)	4
ABS (2 in and larger)	At each story

*All stacks must be supported at their bases.
†Support intervals may not exceed 15 ft.
‡Support intervals may not exceed 30 ft.

FIGURE 5.41 ■ Vertical pipe-support intervals in Zone Two. (*Courtesy of McGraw-Hill*)

Type of vent pipe	Maximum distance between supports (ft)
Lead pipe	Continuous
Cast iron	5*
Galvanized	12
Copper tube (1¼ in)	6
Copper tube (1½ in and larger)	10
ABS	4
PVC	4
Brass	10
Aluminum	10

*Or at every joint.

FIGURE 5.42 ■ Horizontal pipe-support intervals in Zone Three. (*Courtesy of McGraw-Hill*)

Type of vent pipe	Maximum distance between supports (ft)
Lead pipe	4
Cast iron	15
Galvanized	15
Copper tubing	10
ABS	4
PVC	4
Brass	10
Aluminum	15

FIGURE 5.43 ■ Vertical pipe-support intervals in Zone Three. (*Courtesy of McGraw-Hill*)

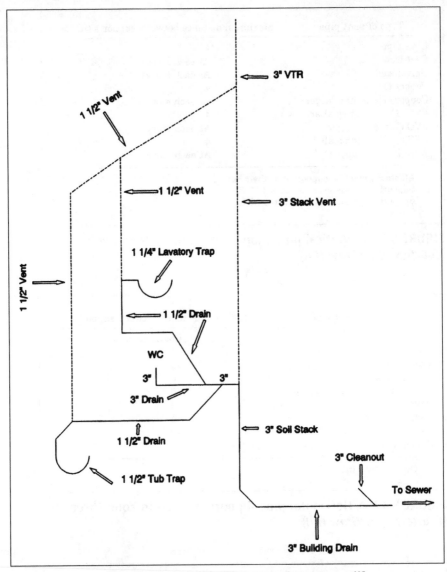

FIGURE 5.44 ▪ DWV riser diagram. (*Courtesy of McGraw-Hill*)

▶ sensible **shortcut**

A rule of thumb for hanger spacing when working with plastic vent piping is to support the pipe at intervals that do not exceed four feet from the center of one support to the center of the next support.

RISER DIAGRAMS

Riser diagrams are often required by code officers prior to any plumbing being installed. Supplying a detailed riser diagram (Fig. 5.44 and Fig. 5.45) is usually a standard part of a permit application. You can also use riser diagrams to help you when sizing a vent system. If

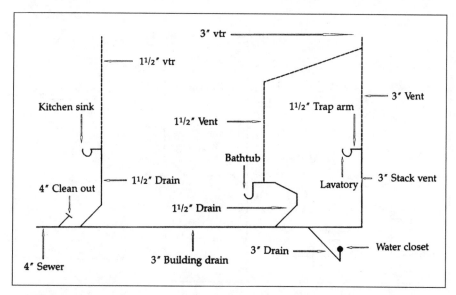

FIGURE 5.45 ■ DWV riser diagram, with size and location of pipes. *(Courtesy of TAB Books,* Home *Plumbing Illustrated, by R. Dodge Woodson, p. 50)*

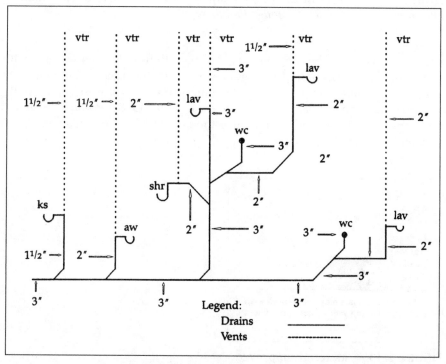

FIGURE 5.46 ■ Poorly designed DWV layouts. *(Courtesy of TAB Books, Home Plumbing Illustrated, by R. Dodge Woodson, p. 55)*

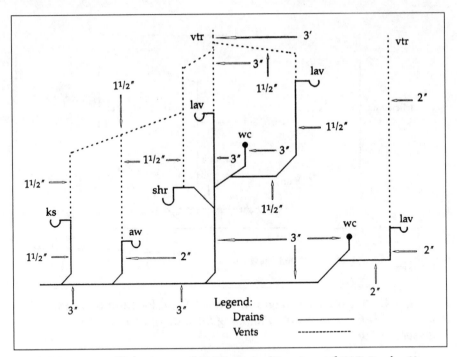

FIGURE 5.47 ■ Efficient use of DWV pipes. (*Courtesy of TAB Books, Home Plumbing Illustrated, by R. Dodge Woodson, p. 55*)

you draw a riser for the job you are working with, the diagram will make it easier for you to label the fixture-unit loads and the sizes of the vents required. Another good use of a riser diagram is to minimize wasted piping. If you draw your piping path on paper, you can spot situations where an alternative plan might be used to minimize the cost of labor and materials (Fig. 5.46 and Fig. 5.47).

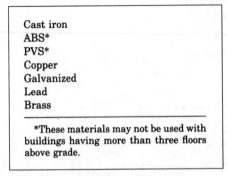

FIGURE 5.48 ■ Materials approved for above-ground vents in Zone One. (*Courtesy of McGraw-Hill*)

CHOOSING MATERIALS

Choosing materials for a venting system is not usually much problem. Most jobs use Schedule-40 plastic pipe for vent pipes. There are, however, other options for vent materials. And, not all codes allow the same types of vent materials. You will also notice from the following tables of approved materials that there can be a difference in approved materials for vents that are installed underground, compared to those installed above ground. We will close out this chapter with tables that indicate what types of materials are allowed within major plumbing codes.

Cast iron
ABS
PVC
Copper
Galvanized
Lead
Aluminum
Borosilicate Glass
Brass

FIGURE 5.49 ■ Materials approved for above-ground vents in Zone Two. (*Courtesy of McGraw-Hill*)

Cast iron
ABS*
PVC*
Copper
Brass
Lead

*These materials may not be used with buildings having more than three floors above grade.

FIGURE 5.51 ■ Materials approved for underground vents in Zone One. (*Courtesy of McGraw-Hill*)

Cast iron
ABS
PVC
Copper
Galvanized
Lead
Aluminum
Brass

FIGURE 5.50 ■ Materials approved for above-ground vents in Zone Three. (*Courtesy of McGraw-Hill*)

Cast iron
ABS
PVC
Copper
Aluminum
Borosilicate glass

FIGURE 5.52 ■ Materials approved for underground vents in Zone Two. (*Courtesy of McGraw-Hill*)

Cast iron
ABS
PVC
Copper

FIGURE 5.53 ■ Materials approved for underground vents in Zone Three. (*Courtesy of McGraw-Hill*)

chapter 6

STORM-WATER CALCULATIONS

Storm-water calculations stump some plumbers. I think that the problem for some plumbers is computing the amount of water accumulated due to structures on roofs. For example, if a roof has an enclosed stairway system, the walls and roof of the stairway have to be factored into the equation for what is required in rainfall drainage. Some plumbers find doing the math for roof drains, rain leaders, and other storm piping to be intimidating. Given the proper charts and tables, the job is not really too difficult.

I used to teach code classes for plumbers who were preparing to take their licensing tests. After teaching the class for a while, I noticed some common elements from class to class. One common thread that seemed to run from class to class was a fear of doing storm-water calculations. I came to expect the classes to be intimidated by what I didn't perceive to be any big deal. Knowing how to size a drainage system for storm water is a requirement for licensing where I live and work, so the people in the class had to address their fears. This, however, was true only of those going for their master's license. Oddly enough, once they were given an example or two of how the work is done, most of them didn't have any problem with their calculations.

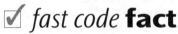

✓ *fast code* **fact**

Journeyman plumbers are not normally required to know how to figure roof drains and major storm-water calculations. This is typically the job of a master plumber. Of course, circumstances vary from location to location, so the process is well worth learning at any level in your plumbing career.

☞ been there **done that**

Don't allow the code requirements to scare you. I remember the first time I had to pipe an island sink. It made me very uncomfortable, even though there was a diagram in the codebook on how to do the job. What may seem daunting when you first look at the code is not necessarily such a mess. Trust in yourself.

APPENDIX A

Rainwater Systems

General

The purpose of this Appendix is to provide drainage from roof areas, courts, and courtyards where it is necessary to collect storm water and deliver to an approved point of disposal not in conflict with other ordinances or regulations.

Part A
Rainwater Systems

A1 Materials:

(a) Rainwater piping placed within the interior of a building or run within a vent or shaft shall be of cast iron, galvanized steel, wrought iron, brass, copper, lead, Schedule 40 ABS DWV, Schedule 40 PVC DWV or other approved materials. ABS and PVC DWV piping installations shall be limited to structures not exceeding three floors above grade. For the purpose of this subsection, the first floor of a building shall be that floor that has fifty (50) percent or more of the exterior wall surface area level with or above finished grade. One (1) additional level that is the first level and not designed for human habitation and used only for vehicle parking, storage, or similar use shall be permitted.

(b) Rainwater piping located on the exterior of a building shall be not less than 26 ga. galvanized sheet metal. When the conductor is connected to a building storm drain or storm sewer, a drain connection shall be extended above the finished grade and jointed at a point protected from injury.

(c) Rainwater piping located underground within a building shall be of service weight cast iron soil pipe, Type DWV copper tube, Schedule 40 ABS DWV, Schedule 40 PVC DWV, extra strength vitrified clay pipe, or other approved materials.

(d) Rainwater piping commencing two (2) feet (.6 m) from the exterior of a building may be of any approved material permitted in the Installation Requirements of this Code.

A1.1

(a) Rainwater piping shall not be used as soil, waste or vent pipes nor shall a soil, waste or vent line be used as a rainwater pipe.

(b) Rainwater piping installed in locations where they may be subjected to damage shall be protected.

FIGURE 6.1 ■ Rainwater code requirements. (*Courtesy of Uniform Plumbing Code*)

I could create some examples for you to work with here, but I won't. Why? Because two of the major codes already offer sample exercises in their codebooks, and the two codes have agreed to allow me to use their examples for this chapter's tutorial. So, what I'm going to do is show you actual excerpts from two codebooks. One of the codes is the Uniform Plumbing Code. The other is the Standard Plumbing Code, or as some people call it, the Southern Plumbing Code. I will let you look over the examples, one at a time, and then I will comment on them, pointing out some of the areas that may appear a little tricky. Let's start with the example provided in the Uniform Plumbing Code. Please refer to Figures 6.1 through 6.8 for code requirements and a sizing example for rainwater systems. I want you to keep in mind that books age

(c)   Roof drains, overflow drains, and rainwater piping installed within the construction of the building shall be tested in conformity with the provisions of this Code for testing drain, waste, and vent systems.

Part B
Roof Drains

A 2   Materials: Roof drains shall be of cast iron, copper, lead, or other corrosion resisting material.

A 2.1 Strainers:

(a)   Roof drains shall be equipped with strainers extending not less than four (4) inches (101.6 mm) above the surface of the roof immediately adjacent to the drain. Strainers shall have minimum inlet area one and one-half (1½) times the pipe to which it is connected.

(b)   Roof deck strainers for use on sun decks, parking decks, and similar occupied areas may be of an approved flat-surface type which is level with the deck. Such drains shall have an inlet area not less than two (2) times the area of the pipe to which the drain is connected.

(c)   Roof drains passing through the roof into the interior of a building shall be made watertight at the roof line by the use of a suitable flashing material.

Part C
Sizing of Rainwater Piping

A 3.1 Vertical rainwater piping shall be sized in accordance with Fig. 8.3. Figure 8.3 is based upon maximum inches (mm) of rainfall per hour falling upon a given roof area in square feet (m^2). Consult local rainfall figures to determine maximum rainfall per hour.

A 3.2 Vertical Wall Areas. Where vertical walls project above a roof so as to permit storm water to drain to the roof area below the adjacent roof area may be computed from Fig. 8.3 as follows:

(a)   For one (1) wall—add fifty (50) percent of the wall area to the roof area figures.

(b)   For two (2) adjacent walls—add thirty-five (35) percent of the total wall areas.

(c)   Two (2) walls opposite of same heights—add no additional area.

(d)   Two (2) walls opposite of differing heights—add fifty (50) percent of wall area above top of lower wall.

(e)   Walls on three (3) sides—add fifty (50) percent of area of the inner wall below the top of the lowest wall, plus allowance for area of wall above top of lowest wall per (b) and (d).

(f)   Walls on four (4) sides—no allowance for all areas below top of lowest wall—add for areas above top of lowest wall per (a), (b), (d), and (e).

FIGURE 6.2 ■ Rainwater code requirements. (*Courtesy of Uniform Plumbing Code*)

and the illustrations here may not be up to speed with your current, local code. Check you own code requirements and use the tables here as examples of how to use what you have.

Now that you've had a chance to look over the illustrations, you may have a solid understanding of how to size a rainwater system. If you do, that's great. But, maybe you have a little confusion that needs to be cleared up. Let me go over a few of the points that some plumbers from my classes have had trouble with. Start by looking at Figure 6.2, part C. In category A 3.2 of Figure 6.2, I want you to look at letter A. The code tells you to figure 50 percent of a single wall for additional rainwater. So, if the wall is 10 feet long and 10 feet tall, its total area would be 100 square feet. This is determined by multiplying the width by the height. In this case, we would add 50 square feet of area to our working numbers to apply to the sizing chart.

Now look at the ruling in letter B. It says that if you have two adjacent walls, you must add 35-percent of their combined area to the equation.

Sizing of Roof Drains and Rainwater Piping for Varying Rainfall Quantities are Horizontal Projected Roof Areas in Square Feet

Rain fall in inches	Size of drain or leader in inches*					
	2	3	4	5	6	8
1	2880	8800	18400	34600	54000	116000
2	1440	4400	9200	17300	27000	58000
3	960	2930	6130	11530	17995	38660
4	720	2200	4600	8650	13500	29000
5	575	1760	3680	6920	10800	23200
6	480	1470	3070	5765	9000	19315
7	410	1260	2630	4945	7715	16570
8	360	1100	2300	4325	6750	14500
9	320	980	2045	3845	6000	12890
10	290	880	1840	3460	5400	11600
11	260	800	1675	3145	4910	10545
12	240	730	1530	2880	4500	9660

Sizing of Roof Drains and Rainwater Piping for Varying Rainfall Quantities are Horizontal Projected Roof Areas in meters^2

Rain fall in mm	Size of drain or leader in millimeters*					
	50.8	76.2	101.6	127	152.4	203.2
25.4	267.6	817.5	1709.4	3214.3	5016.6	10776.4
50.8	133.8	408.8	854.7	1607.2	2508.3	5388.2
76.2	89.2	272.2	569.5	1071.1	1671.7	3591.5
101.6	66.9	204.4	427.3	803.6	1254.2	2694.1
127	53.4	163.5	341.8	642.9	1003.3	2155.3
152.4	44.6	136.6	285.2	535.6	836.1	1794.4
177.8	38.1	117.1	244.3	459.4	716.7	1539.4
203.2	33.4	102.2	213.7	401.8	627.1	1347.1
228.6	29.7	91	190	357.2	557.4	1197.5
254	26.9	81.8	170.9	321.4	501.7	1077.6
279.4	24.2	74.3	155.6	292.2	456.1	979.6
304.8	22.3	67.8	142.1	267.6	418.1	897.4

*Round, square, or rectangular rainwater pipe may be used and are considered equivalent when enclosing a scribed circle equivalent to the leader diameter.

FIGURE 6.3 ■ Rainwater sizing tables. (*Courtesy of Uniform Plumbing Code*)

Assuming that each wall was 10 feet by 10 feet, we would have a total of 200 square feet. 35 percent of 200 square feet is 70 square feet. See how easy this is? In the rulings identified by the letter C, you can see that no additional square footage is added when you have two walls that are opposite of each other and that are the same size. But, letter D offers another ruling. Assume that you have two walls opposite of each other. One of the walls is 10 feet by 10 feet. The other is 10 feet by 15 feet. How much area do you add? One wall is 5 feet taller than the other and 10 feet wide. This amounts to a total area of 50 square feet in differing size for computation purposes. Now all you have to do is divide the difference in half for your working number, which in this case would be 25 square feet. If you pay attention, the code does most of the work for you.

A 3.3 Horizontal Rainwater Piping. The size of a building rainwater piping or any of its horizontal branches shall be sized in accordance with Figs. 8.5 and 8.6 (based upon maximum roof areas to be drained).

Example: Figs. 8.5 and 8.6
Roof Area—5900 sq. ft. (548.1 m²)
Max. Rainfall/hr.—5 inches (127 mm)
Pipe Laid at ¼″ (20.9 mm/m) slope
Find roof area in column under 5″ (127 mm) rainfall (6040 sq. ft. (561.1 m²) is closest), read 6″ (152.4 mm) as size of piping in left hand column.

A 3.4 Roof Gutter. The size of semi-circular roof gutters shall be based on the maximum roof area, in accordance with Figs. 8.7 and 8.8.

Example: Figs. 8.7 and 8.8
Roof Area—2000 sq. ft. (186 m²)
Max. Rainfall/hr.—4″ (101.6 mm)
Gutter set at ⅛″ (10.4 mm/m) slope
Find roof area in column under 4″ (101.6 mm) rainfall 1950 sq. ft. (181.4 m²) is closest), read 7″ (177.8 mm) diameter gutter in left hand column.

A 3.5 If the rainfall is more or less than those shown in Figs. 8.5–8.8, then adjust the figures in the 2″ (50.8 mm) rainfall column by multiplying by two (2) and dividing by the maximum rate of rainfall in inches/hr. (mm/hour).

Example: In Figs. 8.5 and 8.6 with an ⅛″ (10.4 mm/m) slope and an 8″ (203.2 mm) rainfall, find the number of square feet (m²) a 4″ (101.6 mm) pipe will carry.

$$\frac{2 \times 3760}{8} = 940 \text{ sq. ft. } (87.4 \text{ m}^2)$$

FIGURE 6.4 ■ Rainwater sizing example. (*Courtesy of Uniform Plumbing Code*)

Size of Horizontal Rainwater Piping

Size of pipe in inches ⅛"	Maximum rainfall in inches per hour				
	2	3	4	5	6
3	1644	1096	822	657	548
4	3760	2506	1800	1504	1253
5	6680	4453	3340	2672	2227
6	10700	7133	5350	4280	3566
8	23000	15330	11500	9200	7600
10	41400	27600	20700	16580	13800
12	66600	44400	33300	26650	22200
15	109000	72800	59500	47600	39650

Size of pipe in inches ¼" slope	Maximum rainfall in inches per hour				
	2	3	4	5	6
3	2320	1546	1160	928	773
4	5300	3533	2650	2120	1766
5	9440	6293	4720	3776	3146
6	15100	10066	7550	6040	5033
8	32600	21733	16300	13040	10866
10	58400	38950	29200	23350	19450
12	94000	62600	47000	37600	31350
15	168000	112000	84000	67250	56000

Size of pipe in inches ½" slope	Maximum rainfall in inches per hour				
	2	3	4	5	6
3	3288	2295	1644	1310	1096
4	7520	5010	3760	3010	2500
5	13360	8900	6680	5320	4450
6	21400	13700	10700	8580	7140
8	46000	30650	23000	18400	15320
10	85800	55200	41400	33150	27600
12	133200	88800	66600	53200	44400
15	238000	158800	119000	95300	79250

FIGURE 6.5 ■ Rainwater sizing tables. (*Courtesy of Uniform Plumbing Code*)

Size of Horizontal Rainwater Piping					
Size of pipe in mm	Maximum rainfall in millimeters per hour				
10.4 mm/m slope	50.8	76.2	101.6	127	152.4
76.2	152.7	101.8	76.4	61	50.9
101.6	349.3	232.8	174.7	139.7	116.4
127	620.6	413.7	310.3	248.2	206.9
152.4	994	662.7	497	397.6	331.3
203.2	2136.7	1424.2	1068.4	854.7	706
254	3846.1	2564	1923	1540.3	1282
279.4	6187.1	4124.8	3093.6	2475.8	2062.4
381	10126.1	6763.1	5527.6	4422	3683.5

Size of pipe in mm	Maximum rainfall in millimeters per hour				
20.9 mm/m slope	50.8	76.2	101.6	127	152.4
76.2	215.5	143.6	107.8	86.2	71.8
101.6	492.4	328.2	246.2	197	164.1
127	877	584.1	438.5	350.8	292.3
152.4	1402.8	935.1	701.4	561.1	467.6
203.2	3028.5	2019	1514.3	1211.4	1009.5
254	5425.4	3618.5	2712.7	2169.2	1806.9
304.8	8732.6	5815.5	4366.3	3493	2912.4
381	15607.2	10404.8	7803.6	6247.5	5205.4

Size of pipe in mm	Maximum rainfall in millimeters per hour				
41.7 mm/m slope	50.8	76.2	101.6	127	152.4
76.2	305.5	213.2	152.7	121.7	101.8
101.6	698.6	465.4	349.3	279.6	232.3
127	1241.1	826.8	620.6	494.2	413.4
152.4	1988.1	1272.3	994	797.1	663.3
203.2	4274.4	2847.4	2136.7	1709.4	1423.2
254	7692.1	5128.1	3846.1	3079.6	2564
304.8	12374.3	8249.5	6187.1	4942.3	4124.8
381	22110.2	14752.5	11055.1	8853.4	7362.3

FIGURE 6.6 ■ Rainwater sizing tables (metric). (*Courtesy of Uniform Plumbing Code*)

Size of Gutters

Diameter of gutter 1/16" slope	Maximum rainfall in inches per hour				
	2	3	4	5	6
3	340	226	170	136	113
4	720	480	360	288	240
5	1250	834	625	500	416
6	1920	1280	960	768	640
7	2760	1840	1380	1100	918
8	3980	2655	1990	1590	1325
10	7200	4800	3600	2880	2400

Diameter of gutter 1/8" slope	Maximum rainfall in inches per hour				
	2	3	4	5	6
3	480	320	240	192	160
4	1020	681	510	408	340
5	1760	1172	880	704	587
6	2720	1815	1360	1085	905
7	3900	2600	1950	1560	1300
8	5600	3740	2800	2240	1870
10	10200	6800	5100	4080	3400

Diameter of gutter 1/4" slope	Maximum rainfall in inches per hour				
	2	3	4	5	6
3	680	454	340	272	226
4	1440	960	720	576	480
5	2500	1668	1250	1000	834
6	3840	2560	1920	1536	1280
7	5520	3680	2760	2205	1840
8	7960	5310	3980	3180	2655
10	14400	9600	7200	5750	4800

Diameter of gutter 1/2" slope	Maximum rainfall in inches per hour				
	2	3	4	5	6
3	960	640	480	384	320
4	2040	1360	1020	816	680
5	3540	2360	1770	1415	1180
6	5540	3695	2770	2220	1850
7	7800	5200	3900	3120	2600
8	11200	7460	5600	4480	3730
10	20000	13330	10000	8000	6660

FIGURE 6.7 ■ Gutter sizing tables. (*Courtesy of Uniform Plumbing Code*)

Size of Gutters					
Diameter of gutter 5.2 mm/m slope	Maximum rainfall in millimeters per hour				
	50.8	76.2	101.6	127	152.4
76.2	31.6	21	15.8	12.6	10.5
101.6	66.9	44.6	33.4	26.8	22.3
127	116.1	77.5	58.1	46.5	38.7
152.4	178.4	119.1	89.2	71.4	59.5
177.8	256.4	170.9	128.2	102.2	85.3
203.2	369.7	246.7	184.9	147.7	123.1
254	668.9	445.9	334.4	267.6	223
Diameter of gutter 10.4 mm/m slope	Maximum rainfall in millimeters per hour				
	50.8	76.2	101.6	127	152.4
76.2	44.6	29.7	22.3	17.8	14.9
101.6	94.8	63.3	47.4	37.9	31.6
127	163.5	108.9	81.8	65.4	54.5
152.4	252.7	168.6	126.3	100.8	84.1
177.8	362.3	241.5	181.2	144.9	120.8
203.2	520.2	347.5	260.1	208.1	173.7
254	947.6	631.7	473.8	379	315.9
Diameter of gutter 20.9 mm/m slope	Maximum rainfall in millimeters per hour				
	50.8	76.2	101.6	127	152.4
76.2	63.2	42.2	31.6	25.3	21
101.6	133.8	89.2	66.9	53.5	44.6
127	232.3	155	116.1	92.9	77.5
152.4	356.7	237.8	178.4	142.7	118.9
177.8	512.8	341.9	256.4	204.9	170.9
203.2	739.5	493.3	369.7	295.4	246.7
254	133.8	891.8	668.9	534.2	445.9
Diameter of gutter 41.7 mm/m slope	Maximum rainfall in millimeters per hour				
	50.8	76.2	101.6	127	152.4
76.2	89.2	59.5	44.6	35.7	29.7
101.6	189.5	126.3	94.8	75.8	63.2
127	328.9	219.2	164.4	131.5	109.6
152.4	514.7	343.3	257.3	206.2	171.9
177.8	724.6	483.1	362.3	289.9	241.4
203.2	1040.5	693	520.2	416.2	346.5
254	1858	1238.4	929	743.2	618.7

FIGURE 6.8 ■ Gutter sizing tables (metric). (*Courtesy of Uniform Plumbing Code*)

Now, let's look at the example given by the Standard Plumbing Code (Figs. 6.9 to 6.15). Some plumbing codes have recently joined forces to create a cohesive code. Most of this book is based on the International Plumbing Code, but there are others and there are combinations. Keep in mind that every code jurisdiction can create their own amendments to the code, so you must refer to your local, enforceable code to be sure that you are on track with local requirements.

Size of Vertical Leaders

Size of leader or conductor[1] (in)	Maximum projected roof area (sq ft)
2	720
2½	1300
3	2200
4	4600
5	8650
6	13,500
8	29,000

1 in = 25.4 mm
1 ft² = 0.0929 m²
Note:
1. The equivalent diameter of square or rectangular leader may be taken as the diameter of that circle which may be inscribed within the cross-sectional area of the leader.

Size of Horizontal Storm Drains

Diameter of drain (in)	Maximum projected roof area for drains of various slopes (sq ft)		
	⅛ in slope	¼ in slope	½ in slope
3	822	1,160	1,644
4	1,880	2,650	3,760
5	3,340	4,720	6,680
6	5,350	7,550	10,700
8	11,500	16,300	23,000
10	20,700	29,200	41,400
12	33,300	47,000	66,600
15	59,500	84,000	119,000

1 in = 25.4 mm
1 ft² = 0.0929 m²

FIGURE 6.9 ■ Storm drain sizing tables. (*Courtesy of Standard Plumbing Code*)

The sizing example you have just seen is a good, step-by-step example of how to size a drainage system for storm water. You've seen actual code examples and rulings, but remember that these codes are subject to change and may not be the codes being used in your area. Consult your local plumbing code for current, applicable code requirements in your region.

SECONDARY (EMERGENCY) ROOF DRAINS

Secondary Drainage Required
Secondary (emergency) roof drains or scuppers shall be provided where the roof perimeter construction extends above the roof in such a manner that water would be entrapped should the primary drains allow buildup for any reason.

Separate Systems Required
Secondary roof drain systems shall have piping and point of discharge separate from the primary system. Discharge shall be above grade in a location which would normally be observed by the building occupants or maintenance personnel.

Maximum Rainfall Rate for Secondary Drains
Secondary (emergency) roof drain systems or scuppers shall be sized based on the flow rate caused by the 100 year 15 minute precipitation as indicated in Fig. 8.12. The flow through the primary system shall not be considered when sizing the secondary roof drain system.

CONVERSION OF ROOF AREA

General
Where roof drainage is connected to a combined sewer, the drainage area may be converted to equivalent fixture unit loads.

Less Than 256 Fixture Units
When the total fixture unit load on the combined drain is less than 256 fixture units, the equivalent drainage area in horizontal projection shall be taken as 1000 sq ft (92.9 m²).

Greater Than 256 Fixture Units
When the total fixture unit load exceeds 256 fixture units, each additional fixture unit shall be considered the equivalent of 3.9 ft² (0.3623 m²) of drainage area.

Rainfall Other Than 4 Inches (102 mm) Per Hour
If the rainfall to be provided for is more or less than 4 inches (102 mm) per hour, the 1,000 sq ft (92.9 m²) equivalent in 1110.2 and the 3.9 sq ft (0.3623 m²) in 1110.3 shall be adjusted by multiplying by 4 and dividing by the rainfall per hour to be provided for.

VALUES FOR CONTINUOUS FLOW

Where there is a continuous or semicontinuous discharge into the building storm drain or building storm sewer, as from a pump, ejector, air conditioning plant, or similar device, each gallon per minute of such discharge shall be computed as being equivalent to 24 sq ft (2.23 m²) of roof area, based upon a 4-inch (102 mm) rainfall.

FIGURE 6.10A ■ Rainwater code requirements. (*Courtesy of Standard Plumbing Code*)

BACKWATER VALVES

Fixture Branches

Backwater valves shall be installed in the branch of the building drain which receives only the discharge from fixtures located within such branch and shall be located below grade.

Material

Backwater valves shall have all bearing parts of corrosion resistant material. Backwater valves shall comply with ANSI/ASME A112.14.1 or CSA B181.1, CSA B181.2.

Seal

Backwater valves shall be so constructed as to insure a mechanical seal against backflow.

Diameter

Backwater valves, when fully opened, shall have a capacity not less than that of the pipes in which they are installed.

Location

Backwater valves shall be so installed to be accessible for service and repair.

APPENDIX REFERENCES

Additional provisions for storm drainage are found in Appendix B-Roof Drain Sizing Method. These provisions are applicable only where specifically included in the adopting ordinance.

FIGURE 6.10B ■ Rainwater code requirements. (*Courtesy of Standard Plumbing Code*)

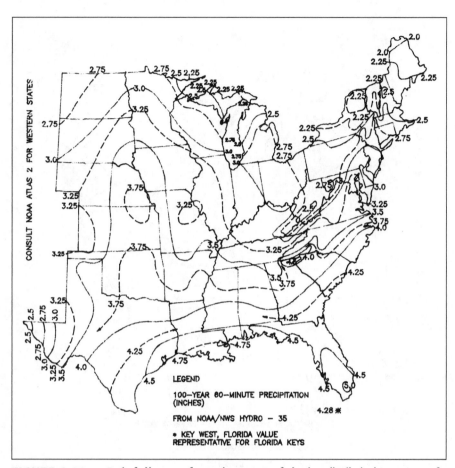

FIGURE 6.11 ■ Rainfall rates for primary roof drains (in/hr). (*Courtesy of Standard Plumbing Code*)

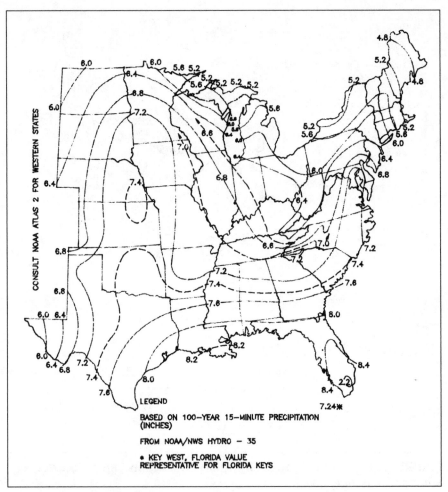

FIGURE 6.12 ■ Rainfall rates for secondary roof drains (in/hr). (*Courtesy of Standard Plumbing Code*)

APPENDIX B
ROOF DRAIN SIZING METHOD

B101 Sizing Example

The following example gives one method of sizing the primary drain system and sizing the scuppers in the parapet walls. This method converts the roof area to an equivalent roof area for a 4-inch rate of rainfall so that Fig. 6.9 can be used as printed.

B101.1 Problem: Given the roof plan in Fig. 6.13 and the site location in Birmingham, Alabama, size the primary roof drain system and size the scuppers, denoting the required head of water above the scupper for the structural engineer.

Note: For the purposes of this appendix the following metric conversions are applicable:

1 in = 25.4 mm

1 ft = 305 mm

1 ft² = 0.0929 m²

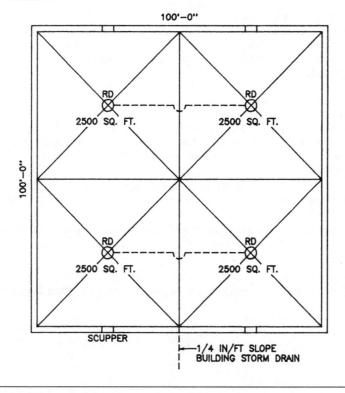

FIGURE 6.13 ■ Example of a roof plan. (*Courtesy of Standard Plumbing Code*)

B101.2 Solution:

Step 1. From Fig. 6.11 the 100 year 60 minute precipitation is 3.75 inches per hour.

Step 2. Each vertical drain must carry 2,500 sq ft of roof area at 3.75 inches per hour of rainfall. To convert to an area for a 4 inch per hour rainfall to enter Fig. 6.9 do this:

2,500 × 3.75 ÷ 4 = 2,344 sq ft. Enter Fig. 6.9 until you find a diameter pipe that will carry 2,344 sq ft. A minimum 4-inch vertical drain is required.

Step 3. Horizontal Drain
2,500 sq ft
To convert to an area for use in Fig. 6.9 do this:

2,500 × 3.75 ÷ 4 = 2,344 sq ft

Enter Fig. 6.9 until you find a diameter pipe that will carry 2,344 sq ft. A minimum 4-inch diameter pipe with a ¼ inch per foot slope will carry 2,650 sq ft. A minimum 4-inch diameter drain on a ¼ inch per foot slope is required.

Step 4. Horizontal Drain
5,000 sq ft
To convert to an area for use in Fig. 6.9 do this:

5,000 × 3.75 ÷ 4 = 4,688 sq ft.

Enter Table 1108.2 until you find a diameter pipe that will carry 4,688 sq ft. A 5-inch diameter pipe with a ¼ inch per foot slope will carry 4,720 sq ft. A minimum 5-inch diameter drain on a ¼ inch per foot slope is required.

Step 5. Horizontal Drain
10,000 sq ft
To convert to an area for use in Fig. 6.9 do this:

10,000 × 3.75 ÷ 4 = 9,375 sq ft

Enter Fig. 6.9 until you find a diameter pipe that will carry 9,375 sq ft. An 8-inch diameter pipe on 1.4 inch per foot slope will carry 16,300 sq ft but a 6-inch will carry only 7,550 sq ft, therefore, use an 8-inch diameter drain on a ¼ inch per foot slope.

Step 6. From Fig. 6.12 the rate caused by a 100 year 15 minute precipitation is 7.2 inches per hour. The scuppers must be sized to carry the flow caused by a rain fall rate of 7.2 inches per hour.

Step 7. Each scupper is draining 2,500 sq ft of roof area. To convert this roof area to an area for use with Fig. 6.15 do this:

2,500 × 7.2 ÷ 4 = 4,500 sq ft

FIGURE 6.14 ▪ Rainwater sizing example. (*Courtesy of Standard Plumbing Code*)

Enter Fig. 6.15 to find a length (see Fig. 6.15) and head that will carry 4,500 sq ft or more.

From Fig. 6.15 a 12-inch wide weir with a 4-inch head carries 6,460 sq ft.

Use 12-inch wide × 5-inch high scuppers at four locations.

A height of 5 inches is needed to assure an open area above the 4-inch head.

Step 8. Notify the structural engineer that the design of the roof structure must account for a height of water to the scupper entrance elevation plus 4 inches for the required head to cause design flow.

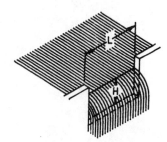

Head (H)	Length (L) of weir (inches)						
inches	4	6	8	12	16	20	24
1	273	418	562	851	1,139	1,427	1,715
2	734	1,141	1,549	2,365	3,180	3,996	4,813
3	1,274	2,023	2,772	4,270	5,768	7,267	8,766
4	1,845	2,999	4,152	6,460	8,766	11,073	13,381
6	2,966	5,087	7,204	11,442	15,680	19,918	24,160

Note:
Table based on rainfall of 4 inches per hour.

FIGURE 6.15 ▪ Scupper sizing table roof area (sq ft.). (*Courtesy of Standard Plumbing Code*)

chapter 7

SIZING WATER HEATERS

Sizing water heaters is not a complicated process. It is, however, an important part of most plumbing jobs. Local code requirements call for minimum standards. The minimum facility requirements can be found in any major codebook or local code enforcement office. Since plumbing codes are regional, you will have to check your local code for exact requirements. But, the math that I'm about to show you will work in any location. Some of the numbers might be different, depending on code requirements, but the mathematical procedure will be the same.

When you figure the size of a water heater, remember that the codes offer suggestions and regulations for minimum requirements. The fact that a 40-gallon water heater will pass code may not mean that it is the best size heater for a given job. Use some common sense when sizing water heaters. Skimping on heater size can prove to be frustrating for your customers; no one enjoys running out of hot water.

There are three types of water heaters that we will discuss. Oil-fired water heaters are the least common of the three. Depending upon where you work, you might find that gas-fired water heaters or electric water heaters are the most prevalent in your region. Overall, electric water heaters are more prolific than gas-fired heaters. Regardless of which type of water heaters you will be working with, you will find the following sizing information helpful.

> ▶ *sensible* **shortcut**
>
> A mistake that some plumbing contractors make is installing water heaters that meet minimum code requirements. This is legal, but it may not make for happy customers. Few people enjoy taking cold showers. With dishwashers running, clothes washers running and large families, the size of a water heater can become very important. If you can get by with a 40-gallon water heater, go with a 52-gallon water heater. Upgrade the size of water heaters to meet your customer needs to avoid complaints down the road.

☑ *fast code* **fact**

Most codes require electric water heaters to have a disconnect box near the appliance and a number 10 electrical wire. In years past, a number 12 wire was acceptable, but this is rarely the case these days. Check your local code for electrical requirements and don't replace or install a water heater that is not in compliance with current code requirements.

ELEMENTS OF SIZING

Elements of sizing are something that you must understand, so let's discuss what they are. The first element is the number of bathrooms in a home or building. When we move to the sizing charts in this chapter, you will see three different formats. This is due to the number of bathrooms. An additional element is the number of bedrooms found in a home. The number of bedrooms is very important. Of course, the storage size of a water heater is a key element. Other elements are the recovery rate, the draw, and the input in either British thermal units per hour (Btuh) or kilowatts (KW). I have prepared some sizing tables for you to use that will make your sizing efforts very easy. Let's look at each table and do a few simple sizing examples to make sure you understand how to use the charts effectively.

HOMES WITH 1 TO 1½ BATHROOMS

We will start our sizing exercises with homes where less than two bathrooms are present. You will see tables for gas-fired, electric, and oil-fired water heaters. The number of bedrooms in our sample homes can range from one to three. You will have to use the chart to size a water heater for the examples given. Let's start with a gas-fired water heater. The house we will size it for will have two bedrooms and one bathroom. What size water heater is needed (Fig. 7.1)?

All you have to do is scan the table for the answer to sizing question. Look under the heading for two bedrooms and run down to the column that lists storage. You will see that a 30-gallon water heater is the minimum size recommended for the application. You will also note that the water heater will recover fully in one hour. Personally, I'd probably up the size of the

Number of bedrooms	1	2	3
Storage capacity (gallons)	20	30	30
Input in Btuh	27,000	36,000	36,000
Draw (gallons per hour)	43	60	60
Recovery (gallons per hour)	23	30	30

FIGURE 7.1 ■ Water heating sizing table for gas heaters (minimum recommendations). Assume less than two full bathrooms.

Number of bedrooms	1	2	3
Storage capacity (gallons)	20	30	30
Input in Btuh	2.5 KW	3.5 KW	4.5 KW
Draw (gallons per hour)	30	44	58
Recovery (gallons per hour)	10	14	18

FIGURE 7.2 ■ Water heating sizing table for electric heaters (minimum recommendations). Assume less than two full bathrooms.

Number of bedrooms	1	2	3
Storage capacity (gallons)	30	30	30
Input in Btuh	70,000	70,000	70,000
Draw (gallons per hour)	89	89	89
Recovery (gallons per hour)	59	59	59

FIGURE 7.3 ■ Water heating sizing table for oil-fired heaters (minimum recommendations). Assume less than two full bathrooms.

heater to 40 gallons, but by code in my region, a 30-gallon tank is all that would be required.

Now, suppose we had the same house but wanted to put an electric water heater in it? What size would we use? Refer to the table in Figure 7.2 to find your answer. In this case, the storage capacity for an electric heater is the same as that required of a gas-fired heater. A 30-gallon tank is all that is needed. But, look at the recovery rate for the electric heater. It's about half as good as the recovery rate for a gas heater. This could be good reason to upgrade the heater to something larger or more powerful.

Let's consider an oil-fired water heater. The basic table (Fig. 7.3) is the same, in terms of use. Again, using the same scenario, what size oil-fired heater would be needed? You will find that a 30-gallon tank is, once again, adequate. Check out the recovery rate. It's great. As you can see, sizing water heaters with the tables provided here is truly easy.

REMAINING TABLES

The remaining tables are different in content, but the procedures for using them are the same. Once you know the number of bedrooms and bathrooms for a dwelling, you can quickly and easily determine the minimum requirements for a water heater. You have just seen how simple the tables are. When you have a water heater to size, just refer to the tables in this chapter (Fig. 7.4, to 7.10) or the tables in your local codebook.

Number of bedrooms	2	3	4	5
Storage capacity (gallons)	30	40	40	50
Input in Btuh	36,000	36,000	38,000	47,000
Draw (gallons per hour)	60	70	72	90
Recovery (gallons per hour)	30	30	32	59

FIGURE 7.4 ■ Water heating sizing table for gas heaters (minimum recommendations). Assume 2 to $2\frac{1}{2}$ bathrooms.

Number of bedrooms	2	3	4	5
Storage capacity (gallons)	40	50	50	66
Input in Btuh	4.5 KW	5.5 KW	5.5 KW	5.5 KW
Draw (gallons per hour)	58	70	72	88
Recovery (gallons per hour)	18	22	22	22

FIGURE 7.5 ■ Water heating sizing table for electric heaters (minimum recommendations). Assume 2 to $2\frac{1}{2}$ bathrooms.

Number of bedrooms	2	3	4	5
Storage capacity (gallons)	30	30	30	30
Input in Btuh	70,000	70,000	70,000	70,000
Draw (gallons per hour)	89	89	89	89
Recovery (gallons per hour)	59	59	59	59

FIGURE 7.6 ■ Water heating sizing table for oil-fired heaters (minimum recommendations). Assume 2 to $2\frac{1}{2}$ bathrooms.

Number of bedrooms	3	4	5	6
Storage capacity (gallons)	40	50	50	50
Input in Btuh	38,000	38,000	47,000	50,000
Draw (gallons per hour)	72	82	90	92
Recovery (gallons per hour)	32	32	40	42

FIGURE 7.7 ▪ Water heating sizing table for gas heaters (minimum recommendations). Assume 3 to 3½ bathrooms.

Number of bedrooms	3	4	5	6
Storage capacity (gallons)	50	66	66	80
Input in Btuh	5.5 KW	5.5 KW	5.5 KW	5.5 KW
Draw (gallons per hour)	72	88	88	102
Recovery (gallons per hour)	22	22	22	22

FIGURE 7.8 ▪ Water heating sizing table for electric heaters (minimum recommendations). Assume 3 to 3½ bathrooms.

Number of bedrooms	3	4	5	6
Storage capacity (gallons)	59	59	59	59
Input in Btuh	70,000	70,000	70,000	70,000
Draw (gallons per hour)	89	89	89	99
Recovery (gallons per hour)	59	59	59	59

FIGURE 7.9 ▪ Water heating sizing table for oil-fired heaters (minimum recommendations). Assume 3 to 3½ bathrooms.

| Column 1 | | Column 2 | |
| Buildings of ordinary tightness | | Buildings of unusually tight construction | |
Condition	Size of opening or duct	Condition	Size of opening or duct
Appliance in unconfined[2] space	May rely on infiltration alone.	Appliance in unconfined[2] space: Obtain combustion air from outdoors or from space freely communicating with outdoors.	Provide two openings, each having 1 sq. in. per 5,000 Btu/h input.
Appliance in confined[4] space 1. All air from inside building	Provide two openings into enclosure each having one square inch per 1,000 Btu/h input freely communicating with other unconfined interior spaces. Minimum 100 sq. in. each opening.	Appliance in confined[4] space: Obtain combustion air from outdoors or from space freely communicating with outdoors.	1. Provide two vertical ducts or plenums: 1 sq. in. per 4,000 Btu/h input each duct or plenum. 2. Provide two horizontal ducts or plenums: 1 sq. in. per 2,000 Btu/h input each duct or plenum. 3. Provide two openings in an exterior wall of the enclosure: each opening 1 sq. in. per 4,000 Btu/h input. 4. Provide one ceiling opening to ventilated attic and one vertical duct to attic: each opening 1 sq. in. per 4,000 Btu/h input. 5. Provide one opening in enclosure ceiling to ventilated attic and one opening in enclosure floor to ventilated crawl space: each opening 1 sq. in. per 4,000 Btu/h input.
2. Part of air from inside building	Provide two openings into enclosure[3] from other freely communicating unconfined[2] interior spaces, each having an area of 100 sq. in. plus one duct or plenum opening to outdoors having an area of 1 sq. in. per 5,000 Btu/h input rating.		
3. All air from outdoors: Obtain from outdoors or from space freely communicating with outdoors.	Use of any of the methods listed for confined space in unusually tight construction as indicated in Column 2.		

[1]For location of opening, see Section 1307(c).
[2]As defined in Section 122.
[3]When the total input rating of appliances in enclosure exceeds 100,000 Btu/h, the area of each opening into the enclosure shall be increased 1 sq. in. for each 1,000 Btu/h over 100,000.
[4]As defined in Section 104(h).

FIGURE 7.10 ■ Size of combustion air openings or ducts for gas-or liquid-burning water heaters. *(Courtesy of Uniform Plumbing Code)*

chapter 8

WATER PUMPS

S ome plumbers work their entire careers without ever having to know
anything about water pumps. Other plumbers deal with pumps on a frequent basis. The difference is where the plumbers work. I've never
worked in New York City, but I suppose there are not many water pumps to
be installed or serviced. But where I live, in Maine, there are more homes
served by private water wells than you can shake a stick at. When I lived in
Virginia, there were plenty of water pumps, too. Some of the pumps are jet
pumps and others are submersible pumps. The two are very different, even
though they do the same job.

Jet pumps are at their best when used in conjunction with shallow wells,
with depths of say 25 feet or less. Two-pipe jet pumps can be used with deep
wells, but a submersible pump is usually a better option for deep wells. Sizing
water pumps and pressure tanks is routine for some plumbers and foreign to
others. This chapter is going to give you plenty of data to use when working
with pump systems.

The illustrations I have to offer you in this chapter are detailed and self-
explanatory. I believe that you will be able to use this chapter as a quick-reference guide to most of your pump questions. Look over the following illustrations and you will find data on jet pumps, submersible pumps, and pressure
tanks. The data will prove very helpful if you become involved with the installation, sizing, or repair of water pumps (Figs. 8.1 to 8.37).

This check list is intended to help in making reliable submersible pump installations. Other data for specific pumps may be needed.

1. Motor Inspection

___ A. Verify that the model, HP or KW, voltage, phase and hertz on the motor nameplate match the installation requirements. Consider any special corrosion resistance required.

___ B. Check that the motor lead assembly is tight in the motor and that the motor and lead are not damaged.

___ C. Test insulation resistance using a 500 or 1000 volt DC megohmmeter, from each lead wire to the motor frame. Resistance should be at least 20 megohms, motor only, no cable.

___ D. Keep a record of motor model number, HP or KW, voltage, date code and serial number.

2. Pump Inspection

___ A. Check that the pump rating matches the motor, and that it is not damaged.

___ B. Verify that the pump shaft turns freely.

3. Pump/Motor Assembly

___ A. If not yet assembled, check that pump and motor mounting faces are free from dirt and uneven paint thickness.

___ B. Assemble the pump and motor together so their mounting faces are in contact, then tighten assembly bolts or nuts evenly to manufacturer specifications. If it is visible, check that the pump shaft is raised slightly by assembly to the motor, confirming impeller running clearance.

___ C. If accessible, check that the pump shaft turns freely.

___ D. Assemble the pump lead guard over the motor leads. Do not cut or pinch lead wire during assembly or handling of the pump during installation.

4. Power Supply and Controls

___ A. Verify that the power supply voltage, hertz, and KVA capacity match motor requirements.

___ B. Use a matching control box with each single phase three wire motor.

___ C. Check that the electrical installation and controls meet all safety regulations and match the motor requirements, including fuse or circuit breaker size and motor overload protection. Connect all metal plumbing and electrical enclosures to the power supply ground to prevent shock hazard. Comply with National and local codes.

5. Lightning and Surge Protection

___ A. Use properly rated surge (lightning) arrestors on all submersible pump installations unless the installation is operated directly from an individual generator and/or is not exposed to surges. Motors 5HP and smaller which are marked "Equipped with Lightning Arrestors" contain internal arrestors.

___ B. Ground all above ground arrestors with copper wire directly to the motor frame, or to metal drop pipe or casing which reaches below the well pumping level. Connecting to a ground rod does not provide good surge protection.

6. Electrical Cable

___ A. Use cable suitable for use in water, sized to carry the motor current without overheating in water and in air, and complying with local regulations. To maintain adequate voltage at the motor, use lengths no longer than specified in the motor manufacturer's cable charts.

___ B. Include a ground wire to the pump if required by codes or surge protection, connected to the power supply ground. Always ground any pump operated outside a drilled well.

7. Well Conditions

___ A. For adequate cooling, motors must have at least the water flow shown on its nameplate. If well conditions and construction do not assure this much water flow will always come from below the motor, use a flow sleeve as shown in the Application, Installation & Maintenance Manual

___ B. If water temperature exceeds 30 degrees C (86 °F), reduce the motor loading or increase the flow rate to prevent overheating, as specified in the Application, Installation & Maintenance Manual.

8. Pump/Motor Installation

___ A. Splice motor leads to supply cable using electrical grade solder or compression connectors, and carefully insulate each splice with watertight tape or adhesive-lined shrink tubing, as shown in motor or pump installation data.

___ B. Support the cable to the delivery pipe every 10 feet (3 meters) with straps or tape strong enough to prevent sagging. Use pads between cable and any metal straps.

___ C. A check valve in the delivery pipe is recommended, even though a pump may be reliable without one. More than one check valve may be required, depending on valve rating and pump setting. Install the lowest check valve below the lowest pumping level of the well, to avoid hydraulic shocks which may damage pipes, valve or motor.

___ D. Assemble all pipe joints as tightly as practical, to prevent unscrewing from motor torque. Recommended torque is at least 10 pound feet per HP (2 meter-KG per KW).

___ E. Set the pump far enough below the lowest pumping level to assure the pump inlet will always have at least the Net Positive Suction Head (NPSH) specified by the pump manufacturer, but at least 10 feet (3 meters) from the bottom of the well to allow for sediment build up.

FIGURE 8.1 ■ Submersible pump installation checklist. (*Courtesy of McGraw-Hill*)

_____ F. Check insulation resistance from dry motor cable ends to ground as the pump is installed, using a 500 or 1000 volt DC megohmmeter. Resistance may drop gradually as more cable enters the water, but any sudden drop indicates possible cable, splice or motor lead damage. Resistance should meet motor manufacturer data.

9. After Installation

_____ A. Check all electrical and water line connections and parts before starting the pump. Make sure water delivery will not wet any electrical parts, and recheck that overload protection in three phase controls meets requirements.

_____ B. Start the pump and check motor amps and pump delivery. If normal, continue to run the pump until delivery is clear. If three phase pump delivery is low, it may be running backward because phase sequence is reversed. Rotation may be reversed (with power off) by interchanging any two motor lead connections to the power supply.

_____ C. Connect three phase motors for current balance within 5% of average, using motor manufacturer instructions. Unbalance over 5% will cause higher motor temperatures and may cause overload trip, vibration, and reduced life.

_____ D. Make sure that starting, running and stopping cause no significant vibration or hydraulic shocks.

_____ E. After at least 15 minutes running, verify that pump output, electrical input, pumping level, and other characteristics are stable and as specified.

Date _____ Filled In By

10. Installation Data

Well Identification _____

Check By _____

Date _____ / _____ / _____

Notes _____

_____

_____

_____

_____

_____

_____

_____

_____

FIGURE 8.1 ■ (_Continued_) Submersible pump installation checklist. (_Courtesy of McGraw-Hill_)

RMA No. _____

INSTALLER'S NAME _____ OWNER'S NAME _____

ADDRESS_____ ADDRESS _____

CITY _____ STATE_____ ZIP_____ CITY _____ STATE_____ ZIP_____

PHONE (____) _____FAX (____) _____ PHONE (____) _____FAX (____) _____

CONTACT NAME _____ CONTACT NAME _____

WELL NAME/ID_____ DATE INSTALLED_____

MOTOR:

Motor No. _____ Date Code _____ HP _____ Voltage _____ Phase _____

PUMP:

Manufacturer _____ Model No. _____ Curve No. _____ Rating: _____ GPM@_____ft. TDH

NPSH Required: _____ ft. NPSH Available:_____ ft. Actual Pump Delivery_____GPM@ _____ PSI

Operating Cycle: _____ON (Min./Hr.) _____ OFF (Min./Hr.) (Circle Min. or Hr. as appropriate)

YOUR NAME _____ DATE _____/_____/_____

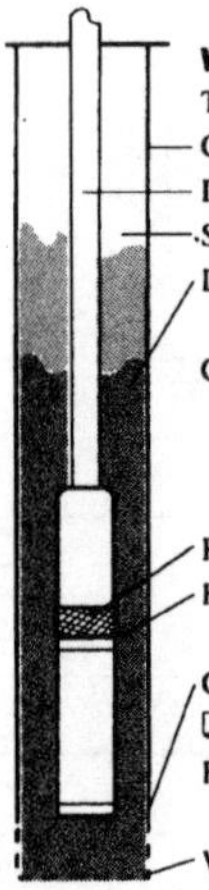

WELL DATA:
Total Dynamic Head _____ft.
Casing Diameter_____in.
Drop Pipe Diameter_____in.
Static Water Level _____ft.
Drawdown (pumping) Water Level_____ft.

Checkvalves at _____&_____&
_____&_____ft.
☐ Solid ☐ Drilled

Pump Inlet Setting _____ft.
Flow Sleeve: ____No____Yes, Dia._____in.

Casing Depth_____ft.
☐ Well Screen ☐ Perforated Casing
From_____to____ft. & _____to____ft.

Well Depth_____ft.

TOP PLUMBING:
Please sketch the plumbing after the well head (check valves, throttling valves, pressure tank, etc.) and indicate the setting of each device.

Form No. 2207 2/94

FIGURE 8.2 ▪ Submersible motor installation record. (*Courtesy of McGraw-Hill*)

Average water requirements for general
service around the home and farm

Each person per day, for all purposes	75 gal.
Each horse, dry cow, or beef animal	12 gal.
Each milking cow	35 gal.
Each hog per day	4 gal.
Each sheep per day	2 gal.
Each 100 chickens per day	4 gal.

Average amount of water required by
various home and yard fixtures

Drinking fountain, continuously flowing	50 to 100 gal. per day
Each shower bath	Up to 30 gal. @ 3–5 gpm
To fill bathtub	30 gal.
To flush toilet	6 gal.
To fill lavatory	2 gal.
To sprinkle 1/4" of water on each 1000 square feet of lawn	160 gal.
Dishwashing machine — per load	7 gal. @ 4 gpm
Automatic washer —per load	Up to 50 gal. @ 4–6 gpm
Regeneration of domestic water softener	50–100 gal.

Average flow rate requirements by
various fixtures
(gpm = gal. per minute; gph = gal. per hour)

Shower	3–5 gpm
Bathtub	3–5 gpm
Toilet	3 gpm
Lavatory	3 gpm
Kitchen sink	2–3 gpm
1/2" hose and nozzle	200 gph
3/4" hose and nozzle	300 gph
Lawn sprinkler	120 gph

FIGURE 8.3 ■ Average water requirements for general service. (*Courtesy of McGraw-Hill*)

	Approx. Gallons Per Day
Each horse	12
Each producing cow	15
Each nonproducing cow	12
Each producing cow with drinking cups	30–40
Each nonproducing cow with drinking cups	20
Each hog	4
Each sheep	2
Each 100 chickens	4–10
Yard fixtures:	
½-inch hose with nozzle	200
¾-inch hose with nozzle	300
Bath houses	10
Camp	
Construction, semipermanent	50
Day (with no meals served)	15
Luxury	100–150
Resorts (day and night, with limited plumbing)	50
Tourists with central bath and toilet facilities	35
Cottages with seasonal occupancy	50
Courts, tourists with individual bath units	50
Clubs	
Country (per resident member)	100
Country (per nonresident member present)	25

FIGURE 8.4 ■ Daily water requirements. (*Courtesy of McGraw-Hill*)

Dwellings
  Luxury   75
  Multiple family, apartments (per resident)   60
  Rooming houses (per resident)   50
  Single family   75
Estates   100–150
Factories (gallons/person/shift)   15–35
Institutions other than hospitals   75–125
  Hospitals (per bed)   250–400
Laundries, self-serviced (gallons per washing, i.e.,
  per customer)   50
Motels
  With bath and toilet (per bed space)   100
Parks
  Overnight with flush toilets   25
  Trailers with individual bath units   50
Picnic
  With bath houses, showers, and flush toilets   20
  With only toilet facilities (gal./picnicker)   10
Restaurants with toilet facilities (per patron)   10
  Without toilet facilities (per patron)   3
  With bars and cocktail lounge (additional quantity)   2
Schools
  Boarding   50–70
  Day with cafeteria, gymnasiums and showers   25
  Day with cafeteria but no gymnasiums or showers   20
Service stations (per vehicle)   10
Stores (per toilet room)   400
Swimming pools   10
Theaters
  Drive-in (per car space)   5
  Movie (per auditorium seat)   5
Workers
  Construction (semipermanent)   50
  Day (school or offices per shift)   15

Providing an adequate water supply provides for a healthy family and higher production from livestock. Assuming the total daily requirement is calculated to be 1200 gpd (gallons per day), a pump would be selected for a capacity of 10 gpm (gallons per minute) based on the following formula:

$$1200 \text{ gph} \div 2 \text{ equals } 600 \text{ gph (gal. per hr.)}$$

Example: 5 in family @ 75 gpd each person   375
          1¾" hose with nozzle @ 300   300
          10 non-producing cows with cups @ 20   <u>200</u>
          Total 24 hr. req.   875

$$875 \div 2 = 438 \text{ gph or } 7.3 \text{ gpm pump selection}$$

FIGURE 8.4 ■ (*Continued*) Daily water requirements. (*Courtesy of McGraw-Hill*)

Engineering Data
Drop Cable Selection Chart
Single-phase, two or three-wire cable, 60 Hz (service entrance to motor)

Motor Rating		Copper Wire Size									
Volts	HP	14	12	10	8	6	4	3	2	1	0
115	1/3	130	210	340	540	840	1300	1610	1960	2390	2910
	1/2	100	160	250	390	620	960	1190	1460	1780	2160
230	1/3	550	880	1390	2190	3400	5250	6520	7960	9690	11770
	1/2	400	650	1020	1610	2510	3880	4810	5880	7170	8720
	3/4	300	480	760	1200	1870	2890	3580	4370	5330	6470
	1	250	400	630	990	1540	2380	2960	3610	4410	5360
	1.5	190	310	480	770	1200	1870	2320	2850	3500	4280
	2	150	250	390	620	970	1530	1910	2360	2930	3620
	3	120	190	300	470	750	1190	1490	1850	2320	2890
	5	0	110*	180	280	450	710	890	1110	1390	1740
	7.5	0	0	120*	200	310	490	610	750	930	1140
	10	0	0	0	160*	250	390	490	600	750	930
	15	0	0	0	0	170*	270	340	430	530	660

1 foot = .3048 meter

Three-phase, three-wire cable, 60 Hz 200 and 230 volts (service entrance to motor)

Motor Rating		Copper Wire Size (1)												
Volts	HP	14	12	10	8	6	4	3	2	1	0	00	000	0000
200V 60 Hz Three-Phase Three-Wire	1/2	710	1140	1800	2840	4420								
	3/4	510	810	1280	2030	3160								
	1	430	690	1080	1710	2670	4140	5140						
	1.5	310	500	790	1260	1960	3050	3780						
	2	240	390	610	970	1520	2360	2940	3610	4430	5420			
	3	180	290	470	740	1160	1810	2250	2760	3390	4130			
	5	110*	170	280	440	690	1000	1350	1680	2040	2490	3050	3670	4440
	7.5	0	0	200	310	490	770	960	1180	1450	1770	2170	2600	3150
	10	0	0	150*	230	370	570	720	880	1090	1330	1640	1970	2390
	15	0	0	0	160	250	390	490	600	740	910	1110	1340	1630
	20	0	0	0	0	190*	300	380	460	570	700	860	1050	1270
	25	0	0	0	0	0	240*	300	370	460	570	700	840	1030
	30	0	0	0	0	0	200*	250*	310	380	470	580	700	850
230V 60 Hz Three-Phase Three-Wire	1/2	810	1300	2040	3210	4990								
	3/4	590	940	1480	2330	3620								
	1	490	790	1240	1960	3050	4720	5860						
	1.5	360	580	920	1450	2260	3510	4360						
	2	280	450	700	1110	1740	2710	3370	4130	5070	6200			
	3	210	340	540	860	1340	2080	2580	3880	4730				
	5	130*	200	320	510	800	1240	1550	1900	2330	2850	3490	4200	5080
	7.5	0	140*	230	360	570	890	1100	1350	1660	2030	2480	2980	3600
	10	0	0	170*	270	420	660	820	1010	1240	1520	1870	2260	2740
	15	0	0	0	180*	290	450	560	690	850	1040	1280	1540	1860
	20	0	0	0	140*	220*	350	430	530	660	810	990	1200	1450
	25	0	0	0	0	180*	280	350	430	530	650	800	970	1170
	30	0	0	0	0	0	230*	290	350	440	540	660	800	970
460V 60 Hz Three-Phase Three-Wire	1/2	3770	6020	9460										
	3/4	2730	4350	6850										
	1	2300	3670	5770	9070									
	1.5	1700	2710	4270	6730									
	2	1300	2070	3270	5150	8050								
	3	1000	1600	2520	3970	6200								
	5	590	950	1500	2360	3700	5750							
	7.5	420	680	1070	1690	2640	4100	5100	6260	7680				
	10	310	500	790	1250	1960	3050	3800	4680	5750	7050			
	15	0	340*	540	850	1340	2090	2600	3200	3930	4810	5900	7110	
	20	0	0	410	650	1030	1610	2000	2470	3040	3730	4580	5530	
	25	0	0	330*	530	830	1300	1620	1990	2450	3010	3700	4470	5430
	30	0	0	270*	430	680	1070	1330	1640	2030	2490	3060	3700	4500
	40	0	0	0	320*	500*	790	980	1210	1490	1830	2250	2710	3290
	50	0	0	0	0	410*	640	800	980	1210	1480	1810	2190	2650
	60	0	0	0	0	0	540*	670*	830	1020	1250	1540	1850	2240
	75	0	0	0	0	0	440*	550*	680*	840	1030	1260	1520	1850
	100	0	0	0	0	0	0	0	500*	620*	760*	940	1130	1380
	125	0	0	0	0	0	0	0	0	600*	740*	890*	1000	
	150	0	0	0	0	0	0	0	0	0	0	630*	780*	920*
	175	0	0	0	0	0	0	0	0	0	0	0	670*	810*
	200	0	0	0	0	0	0	0	0	0	0	0	590*	710*

Lengths marked * meet the U.S. National Electrical Code ampacity only for **individual** conductor 75°C cable. Only the lengths **without** * meet the code for **jacketed** 75°C cable. Local code requirements may vary.
CAUTION!! Use of wire sizes smaller than determined above **will void warranty**, since low starting voltage and early failure of the unit will result. Larger wire sizes (smaller numbers) may always be used to improve economy of operation.
(1) If aluminum conductor is used, multiply above lengths by 0.61. Maximum allowable length of aluminum wire is considerably shorter than copper wire of same size.

FIGURE 8.5 ■ Drop cable selection chart. (*Courtesy of McGraw-Hill*)

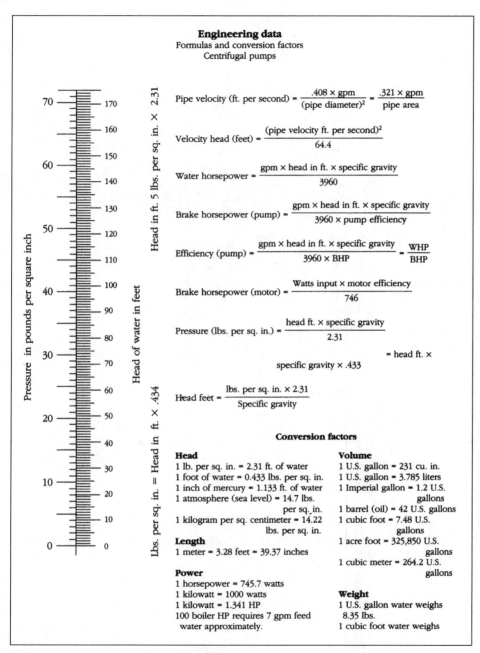

Engineering data
Formulas and conversion factors
Centrifugal pumps

Pipe velocity (ft. per second) = $\dfrac{.408 \times gpm}{(\text{pipe diameter})^2} = \dfrac{.321 \times gpm}{\text{pipe area}}$

Velocity head (feet) = $\dfrac{(\text{pipe velocity ft. per second})^2}{64.4}$

Water horsepower = $\dfrac{gpm \times \text{head in ft.} \times \text{specific gravity}}{3960}$

Brake horsepower (pump) = $\dfrac{gpm \times \text{head in ft.} \times \text{specific gravity}}{3960 \times \text{pump efficiency}}$

Efficiency (pump) = $\dfrac{gpm \times \text{head in ft.} \times \text{specific gravity}}{3960 \times BHP} = \dfrac{WHP}{BHP}$

Brake horsepower (motor) = $\dfrac{\text{Watts input} \times \text{motor efficiency}}{746}$

Pressure (lbs. per sq. in.) = $\dfrac{\text{head ft.} \times \text{specific gravity}}{2.31}$

= head ft. × specific gravity × .433

Head feet = $\dfrac{\text{lbs. per sq. in.} \times 2.31}{\text{Specific gravity}}$

Conversion factors

Head
1 lb. per sq. in. = 2.31 ft. of water
1 foot of water = 0.433 lbs. per sq. in.
1 inch of mercury = 1.133 ft. of water
1 atmosphere (sea level) = 14.7 lbs. per sq. in.
1 kilogram per sq. centimeter = 14.22 lbs. per sq. in.

Length
1 meter = 3.28 feet = 39.37 inches

Power
1 horsepower = 745.7 watts
1 kilowatt = 1000 watts
1 kilowatt = 1.341 HP
100 boiler HP requires 7 gpm feed water approximately.

Volume
1 U.S. gallon = 231 cu. in.
1 U.S. gallon = 3.785 liters
1 Imperial gallon = 1.2 U.S. gallons
1 barrel (oil) = 42 U.S. gallons
1 cubic foot = 7.48 U.S. gallons
1 acre foot = 325,850 U.S. gallons
1 cubic meter = 264.2 U.S. gallons

Weight
1 U.S. gallon water weighs 8.35 lbs.
1 cubic foot water weighs

FIGURE 8.6 ▪ Formulas and conversion factors for centrifugal pumps. (*Courtesy of McGraw-Hill*)

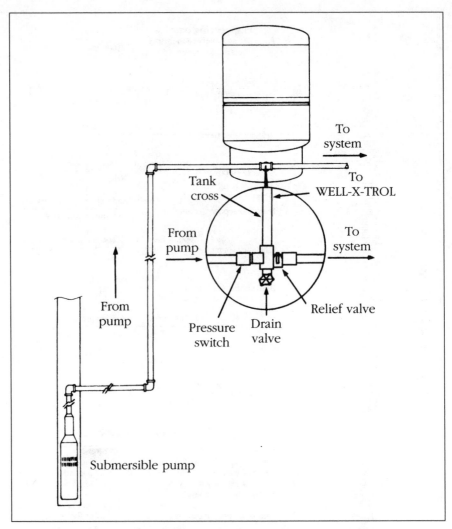

FIGURE 8.7 ■ Pressure tank in use with a submersible pump. (*Courtesy of McGraw-Hill*)

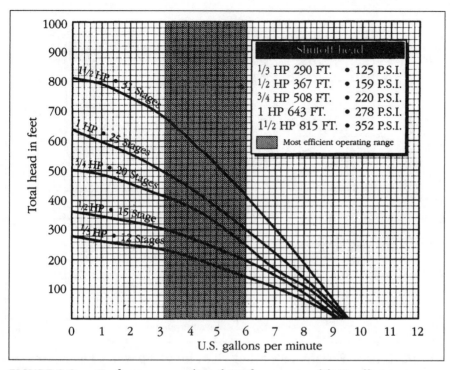

FIGURE 8.8 ▪ Performance rating chart for pump with 5 gallon-per-minute output. (*Courtesy of McGraw-Hill*)

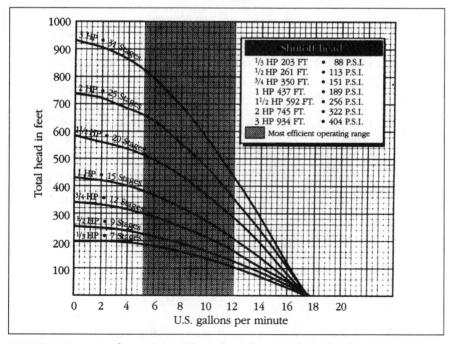

FIGURE 8.9 ▪ Performance rating chart for pump with 10 gallon-per-minute output. (*Courtesy of McGraw-Hill*)

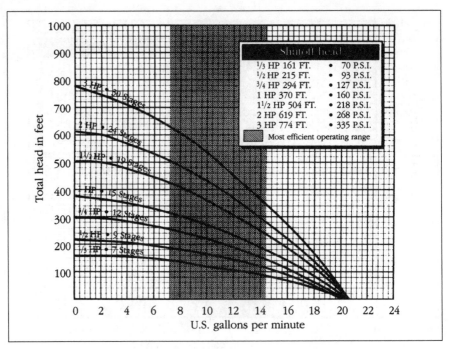

FIGURE 8.10 ■ Performance rating chart for pump with 13 gallon-per-minute output. (*Courtesy of McGraw-Hill*)

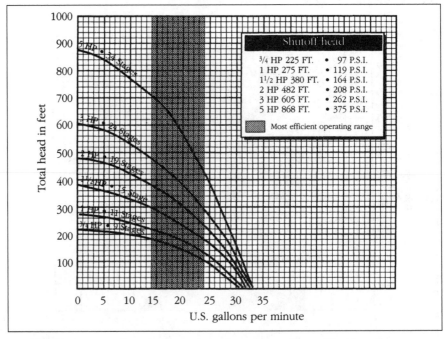

FIGURE 8.11 ■ Performance rating chart for pump with 18 gallon-per-minute output. (*Courtesy of McGraw-Hill*)

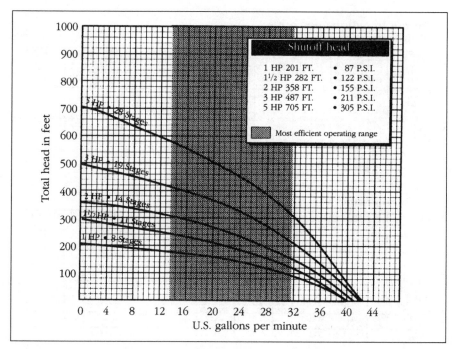

FIGURE 8.12 ■ Performance rating chart for pump with 25 gallon-per-minute output. (*Courtesy of McGraw-Hill*)

Output in gallons per hour
Discharge pressure 0 P.S.I.

	DEPTH	25'	50'	75'	100'	125'	150'	175'	200'	225'	250'	275'	300'	325'	350'	375'	400'	425'	450'	475'
J	1/3 H.P.	540	510	470	425	390	350	310	270	210	125									
	1/2 H.P.	550	520	490	465	440	410	380	350	320	290	255	220	150						
	3/4 H.P.	560	540	520	490	470	445	425	405	385	365	340	325	300	275	245	215	185	145	100
	1 H.P.	565	545	530	510	495	475	460	440	420	400	385	365	350	330	305	290	265	240	220
K	1/3 H.P.	970	890	810	725	630	530	400												
	1/2 H.P.	995	950	865	790	720	635	540	445	325	160									
	3/4 H.P.	1005	960	900	850	800	760	700	635	560	490	410	325	205						
	1 H.P.	1020	975	940	895	860	820	775	725	680	640	590	535	475	420	325	240	120		
L	1/3 H.P.	1100	970	840	680	480	205													
	1/2 H.P.	1120	1025	930	820	710	570	420	210											
	3/4 H.P.	1150	1085	1020	955	875	800	715	620	505	370	200								
	1 H.P.	1170	1115	1060	1005	950	880	820	755	685	605	520	420	300	155					
P	3/4 H.P.	1800	1650	1560	1440	1290	1110	870	600											
	1 H.P.	1830	1725	1650	1560	1440	1305	1140	960	720	450									
M	1 H.P.	2290	2110	1930	1740	1500	1152	720												

Discharge pressure 30 P.S.I.

	DEPTH	25'	50'	75'	100'	125'	150'	175'	200'	225'	250'	275'	300'	325'	350'	375'	400'	425'	450'	475'
J	1/3 H.P.	440	395	360	320	275	230	160												
	1/2 H.P.	475	445	415	385	360	330	295	260	220	175									
	3/4 H.P.	505	475	450	425	410	390	370	350	335	305	280	250	230	190	150	105			
	1 H.P.	515	500	480	460	450	425	400	385	370	355	330	310	295	270	245	220	205	175	145
K	1/3 H.P.	785	695	603	488	333														
	1/2 H.P.	815	730	670	550	470	360													
	3/4 H.P.	875	815	755	705	660	590	505	420	360	240									
	1 H.P.	910	865	820	780	745	695	650	600	555	480	420	335	275						
L	1/3 H.P.	708	495																	
	1/2 H.P.	795	730	595	440															
	3/4 H.P.	985	890	815	733	635	525	393												
	1 H.P.	1015	958	895	833	765	695	608	535	433	320									
P	3/4 H.P.	1470	1320	1140	920	600														
	1 H.P.	1570	1460	1230	1180	950	760													
M	1 H.P.	1778	1540	1194	785															

Discharge pressure 40 P.S.I.

	DEPTH	25'	50'	75'	100'	125'	150'	175'	200'	225'	250'	275'	300'	325'	350'	375'	400'	425'	450'	475'
J	1/3 H.P.	395	365	325	280	215	170	120												
	1/2 H.P.	440	420	390	360	330	300	280	230	180	105									
	3/4 H.P.	480	455	430	410	385	370	355	330	310	285	255	225	195	155	120				
	1 H.P.	500	485	465	445	425	405	390	370	355	335	310	295	265	250	225	205	175	145	115
K	1/3 H.P.	745	650	555	430	240														
	1/2 H.P.	745	670	615	485	365	230													
	3/4 H.P.	825	780	720	660	590	515	445	360	250										
	1 H.P.	875	830	790	745	695	660	585	550	495	440	360	275	170						
L	1/3 H.P.	550	280																	
	1/2 H.P.	745	620	470	280															
	3/4 H.P.	900	820	740	650	540	415	260												
	1 H.P.	960	900	835	775	700	630	520	455	335	205									
P	3/4 H.P.	1320	1230	920	690															
	1 H.P.	1460	1380	1200	1020	790														
M	1 H.P.	1588	1272	852	360															

Discharge pressure 50 P.S.I.

	DEPTH	25'	50'	75'	100'	125'	150'	175'	200'	225'	250'	275'	300'	325'	350'	375'	400'	425'	450'	475'
J	1/3 H.P.	365	325	280	235	180														
	1/2 H.P.	420	390	360	330	300	265	230	175	110										
	3/4 H.P.	460	435	415	390	370	355	330	310	285	260	225	200	160	115					
	1 H.P.	485	470	445	425	405	390	370	355	335	315	295	270	250	230	205	175	145	130	
K	1/3 H.P.	605	475																	
	1/2 H.P.	670	585	480	360	240														
	3/4 H.P.	770	720	660	585	515	445	360	250	120										
	1 H.P.	835	790	745	695	660	610	550	490	430	365	275	170							
L	1/2 H.P.	618	448																	
	3/4 H.P.	830	743	653	545	410														
	1 H.P.	905	843	778	708	630	550	435	340											
P	3/4 H.P.	1170	960	720																
	1 H.P.	1390	1200	1030	810															
M	1 H.P.	1232	876																	

FRICTION LOSSES IN RISER PIPE HAVE NOT BEEN CALCULATED

FIGURE 8.13 ■ Output performance chart for submersible pump. (*Courtesy of McGraw-Hill*)

Model no.	HP	Volts	Impeller material	Pres. switch setting	Suction pipe size	Discharge size	Shipping weight
8130	⅓	115	Plastic	20-40	1¼"	¾"	46 lbs.
8131	⅓	115	Brass	20-40	1¼"	¾"	48 lbs.
8150	½	115/230	Plastic	20-40	1¼"	¾"	48 lbs.
8151	½	115/230	Brass	20-40	1¼"	¾"	50 lbs.
8170	¾	115/230	Plastic	30-50	1¼"	¾"	50 lbs.
8171	¾	115/230	Brass	30-50	1¼"	¾"	52 lbs.
8110	1	115/230	Plastic	30-50	1¼"	¾"	52 lbs.
8111	1	115/230	Brass	30-50	1¼"	¾"	53 lbs.

FIGURE 8.14 ■ Performance ratings for jet pumps. (*Courtesy of McGraw-Hill*)

Model	Model	HP	Volts	Impeller material	Pres. switch setting	Suction pipe size	Twin type drop pipe	Shipping weight
1550	1050	½	115/230	Brass	30-50	1¼"	1" × 1¼"	65 lbs.
1575	1075	¾	115/230	Brass	30-50	1¼"	1" × 1¼"	71 lbs.
1575SW	1075SW	¾	115/230	Brass	30-50	1¼"	1" × 1¼"	66 lbs.
1510	1010	1	115/230	Brass	30-50	1¼"	1" × 1¼"	74 lbs.
1510SW	1010SW	1	115/230	Brass	30-50	1¼"	1" × 1¼"	67 lbs.
1515SW	1015SW	1½	115/230	Brass	30-50	1¼"	1" × 1¼"	72 lbs.

FIGURE 8.15 ■ Performance ratings for multi-stage pumps. *(Courtesy of McGraw-Hill)*

Model no.	HP	Suction lift–ft.	20
8130	⅓	5	755
8131		15	600
		25	345
8150	½	5	935
8151		15	700
		25	395
8170	¾	5	1130
8171		15	865
		25	515
8110	1	5	1605
8111		15	1240
		25	775

FIGURE 8.16 ■ Shallow-well performance chart. (*Courtesy of McGraw-Hill*)

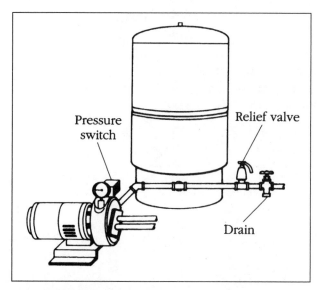

FIGURE 8.17 ■ A typical jet-pump set-up. (*Courtesy of McGraw-Hill*)

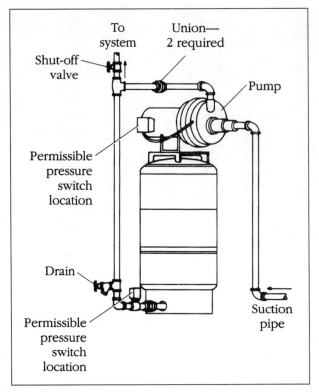

FIGURE 8.18 ■ A jet pump mounted on a pressure tank with a pump bracket. (*Courtesy of McGraw-Hill*)

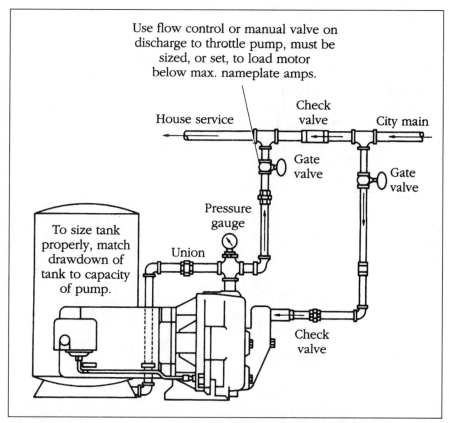

Use flow control or manual valve on
discharge to throttle pump, must be
sized, or set, to load motor
below max. nameplate amps.

House service

Check
valve

City main

Gate
valve

Gate
valve

Pressure
gauge

To size tank
properly, match
drawdown of
tank to capacity
of pump.

Union

Check
valve

FIGURE 8.19 ■ A typical piping arrangement for a jet pump. (*Courtesy of McGraw-Hill*)

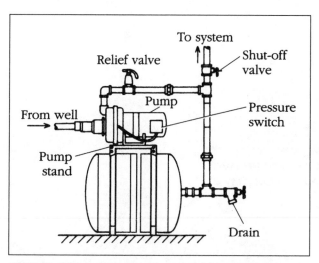

To system

Relief valve

Shut-off
valve

Pump

Pressure
switch

From well

Pump
stand

Drain

FIGURE 8.20 ■ Bracket-mounted jet pump on a horizontal pressure tank. (*Courtesy of McGraw-Hill*)

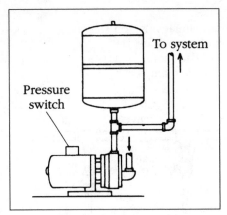

FIGURE 8.21 ■ Small, vertical pressure tank installed above pump. (*Courtesy of McGraw-Hill*)

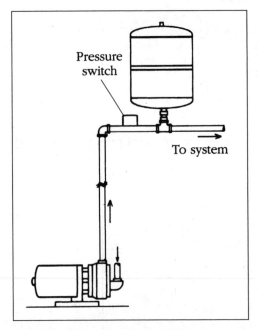

FIGURE 8.22 ■ Small, vertical pressure tank installed above pump. (*Courtesy of McGraw-Hill*)

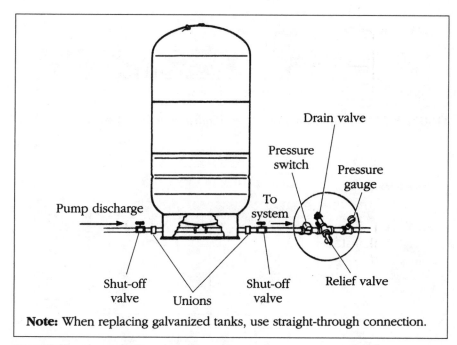

FIGURE 8.23 ■ Stand-type pressure tank with a straight-through method not using a tank fee. (*Courtesy of McGraw-Hill*)

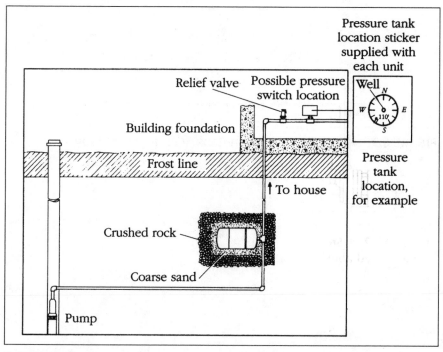

FIGURE 8.24 ■ An underground installation of a pressure tank. (*Courtesy of McGraw-Hill*)

In-line models

Model No.	Dimensions		Total Volume (gals)	Max. Accept. Factor	Drawdown			Shipping Wt.(Vol.) lbs(cu ft)
	Diameter (ins)	Height (ins)			20/40 (gals)	30/50 (gals)	40/60 (gals)	
WX-101	8	12 5/8	2.0	0.45	.7	.6	.5	5 (0.6)

FIGURE 8.25 ■ In-line pressure tank. (*Courtesy of McGraw-Hill*)

Stand models

Model No.	Dimensions		Total Volume (gals)	Max. Accept. Factor	Drawdown			Shipping Wt. (Vol.) lbs (cu ft)
	Diameter (ins)	Height (ins)			20/40 (gals)	30/50 (gals)	40/60 (gals)	
WX-104-S	15 3/8	19 1/4	10.3	1.00	3.8	3.2	2.8	25 (3.0)
WX-201	15 3/8	23 7/8	14.0	0.81	5.1	4.3	3.7	27 (3.8)
WX-202	15 3/8	31 5/8	20.0	0.57	7.3	6.2	5.4	35 (4.9)
WX-203	15 3/8	46 3/8	32.0	0.35	—	9.9	8.6	43 (7.0)
WX-104-LTD	15 3/8	19 1/4	10.3	1.00	3.8	3.2	2.7	23 (3.0)
WX-201-LTD	15 3/8	24	14.0	0.81	5.2	4.3	3.8	25 (3.8)
WX-202-LTD	15 3/8	31 3/4	20.0	0.57	7.4	6.2	5.4	33 (4.9)
WX-203-LTD	15 3/8	46 5/8	35.0	0.35	—	9.9	8.6	43 (7.0)
WX-205	22	29 1/2	34.0	1.00	12.4	10.5	9.1	61 (9.5)
WX-250	22	35 5/8	44.0	0.77	16.3	13.6	11.9	69 (11.0)
WX-251	22	46 3/4	62.0	0.55	22.9	19.2	16.7	92 (13.9)
WX-302	26	47 3/16	86.0	0.54	31.8	26.7	23.2	123 (18.9)
WX-350	26	61 7/8	119.0	0.39	44.0	36.9	32.1	166 (24.5)
WX-302-HC	26	47 3/16	86.0	0.54	—	—	—	125 (18.9)
WX-350-HC	26	61 7/8	119.0	0.39	—	—	—	168 (24.5)

Precharge Pressure for WX-104-S thru WX-203 is 30 PSIG and Sys. Conn. is 1" NPTF.
Precharge Pressure for WX-205 thru WX-350-HC is 38 PSIG and Sys. Conn. is 1 1/4" NPTF.
Maximum Working Pressure is 100 PSIG and Maximum Working Temperature is 200° F.

Note: Drawdown can be affected by various ambient and system conditions, including temperature and pressure.

FIGURE 8.26 ■ Stand-type pressure tank. (*Courtesy of McGraw-Hill*)

7 bar series

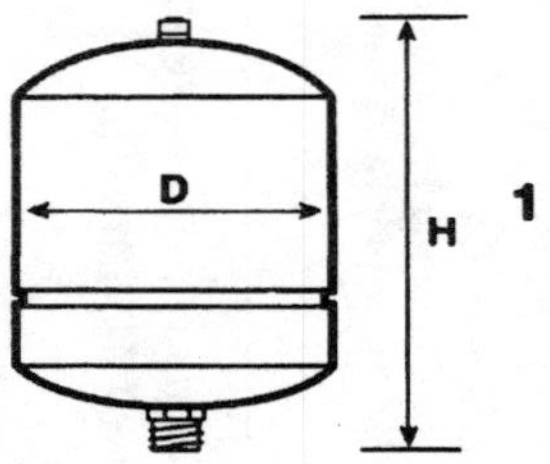

Model No.	Dimensions		Total Volume	1,5/3,0 bar	2,0/3,5 bar	2,5/4,0 bar	System Connection	Precharge Pressure	Shipping Wt./Vol.
	Diameter	Height			Drawdown				
	mm	mm	Ltr	Liter	Liter	Liter	R"	bar	KG / m3
WX 2,6	156	228	2,6	1,0	0,9	0,8	3/4	1,5	1,0 / ,005
WX 4	156	302	4,1	1,5	1,4	1,2	3/4	1,5	1,5 / ,007
WX 8	200	320	8	3,0	2,6	2,4	3/4	1,5	2,3 / ,02
WX 18	280	380	18	6,7	6,0	5,4	3/4	1,5	4,1 / ,03
WX 33	280	630	33	12,4	10,9	9,9	3/4	1,5	6,8 / ,05

FIGURE 8.27 ■ Specifications for in-line pressure tanks. (*Courtesy of McGraw-Hill*)

10 bar series

Model No.	Dimensions Diameter	Height	Total Volume
	mm	mm	Ltr
WL 1855	560	805	80
WL 1856	560	1240	180
WL 1858	560	1700	300
WL 1859	750	1880	600
WL 1860	750	2340	800
WL 1861	1000	1960	1000
WL 1862	1000	2740	1600
WL 1863	1200	2493	2000

Design with stainless steel system connection (V4A)

WL 1935	560	805	80
WL 1936	560	1240	180
WL 1938	560	1700	300
WL 1939	750	1880	600
WL 1940	750	2340	800
WL 1941	1000	1960	1000
WL 1942	1000	2740	1600
WL 1943	1200	2493	2000

16 bar series

WL 1955	560	805	80
WL 1956	560	1240	180
WL 1958	560	1700	300
WL 1959	750	1880	600
WL 1960	750	2340	800
WL 1961	1000	1960	1000
WL 1962	1000	2740	1600
WL 1963	1200	2493	2000

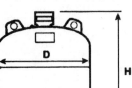
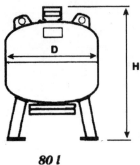

80 l

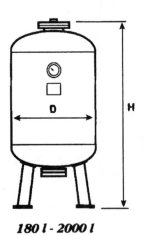

180 l - 2000 l

Maximum Operating Temperature = 90°C.
Horizontal designs and tanks for Operating Pressures of 25 bar are available on request.
Note: Drawdown can be affected by ambient and system conditions, including temperature and pressure.

FIGURE 8.28 ■ Specifications for pressure tanks with replaceable bladder designs. (*Courtesy of McGraw-Hill*)

| 1,5/3,0 bar | 2,0/3,5 bar | 2,5/4,0 bar | System Connection | Precharge Pressure | Shipping Wt./Vol. |
| | Drawdown | | | | |
Liter	Liter	Liter	R"	bar	KG / m3
30	27	24	2	3,5	59 / ,25
68	60	54	2	3,5	83 / ,39
113	99	90	2	3,5	155 / ,53
225	198	180	2	3,5	285 / 1,06
300	264	240	2	3,5	360 / 1,32
375	330	300	3	3,5	400 / 1,96
600	528	480	3	3,5	540 / 2,74
750	660	600	3	3,5	780 / 3,59
30	27	24	2	3,5	59 / ,25
68	60	54	2	3,5	83 / ,39
113	99	90	2	3,5	155 / ,53
225	198	180	2	3,5	285 / 1,06
300	264	240	2	3,5	360 / 1,32
375	330	300	3	3,5	400 / 1,96
600	528	480	3	3,5	540 / 2,74
750	660	600	3	3,5	7890 / 3,59
30	27	24	2	3,5	64 / ,25
68	60	54	2	3,5	102 / ,39
113	99	90	2	3,5	220 / ,53
225	198	180	2	3,5	400 / 1,06
300	264	240	2	3,5	505 / 1,32
375	330	300	3	3,5	560 / 1,96
600	528	480	3	3,5	756 / 2,74
750	660	600	3	3,5	1330 / 3,5

FIGURE 8.28 ■ (*Continued*) Specifications for pressure tanks with replaceable bladder designs. (*Courtesy of McGraw-Hill*)

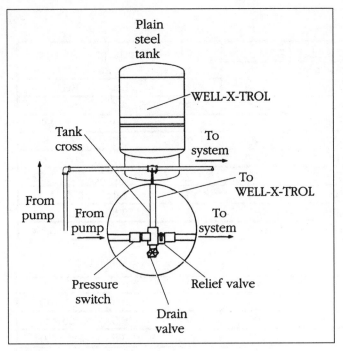

FIGURE 8.29 ■ Detail for a tank-tee set-up. (*Courtesy of McGraw-Hill*)

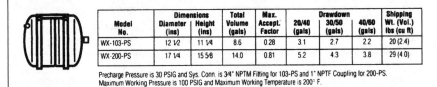

		Dimensions		Total Volume (gals)	Max. Accept. Factor	Drawdown			Shipping Wt. (Vol.) lbs (cu ft)
Model No.		Diameter (ins)	Height (ins)			20/40 (gals)	30/50 (gals)	40/60 (gals)	
WX-103-PS		12 1/2	11 1/4	8.6	0.28	3.1	2.7	2.2	20 (2.4)
WX-200-PS		17 1/4	15 5/8	14.0	0.81	5.2	4.3	3.8	29 (4.0)

Precharge Pressure is 30 PSIG and Sys. Conn. is 3/4" NPTM Fitting for 103-PS and 1" NPTF Coupling for 200-PS.
Maximum Working Pressure is 100 PSIG and Maximum Working Temperature is 200° F.

FIGURE 8.30 ■ Pump-stand type of pressure tank. (*Courtesy of McGraw-Hill*)

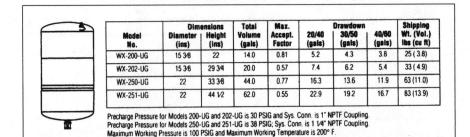

		Dimensions		Total Volume (gals)	Max. Accept. Factor	Drawdown			Shipping Wt. (Vol.) lbs (cu ft)
Model No.		Diameter (ins)	Height (ins)			20/40 (gals)	30/50 (gals)	40/60 (gals)	
WX-200-UG		15 3/8	22	14.0	0.81	5.2	4.3	3.8	25 (3.8)
WX-202-UG		15 3/8	29 3/4	20.0	0.57	7.4	6.2	5.4	33 (4.9)
WX-250-UG		22	33 3/8	44.0	0.77	16.3	13.6	11.9	63 (11.0)
WX-251-UG		22	44 1/2	62.0	0.55	22.9	19.2	16.7	83 (13.9)

Precharge Pressure for Models 200-UG and 202-UG is 30 PSIG and Sys. Conn. is 1" NPTF Coupling.
Precharge Pressure for Models 250-UG and 251-UG is 38 PSIG; Sys. Conn. is 1 1/4" NPTF Coupling.
Maximum Working Pressure is 100 PSIG and Maximum Working Temperature is 200° F.

FIGURE 8.31 ■ Underground pressure tank specifications. (*Courtesy of McGraw-Hill*)

1.
WELL-X-TROL has a sealed-in air chamber that is pre-pressurized before it leaves our factory. Air and water do not mix eliminating any chance of "waterlogging" through loss of air to system water.

2.
When the pump starts, water enters the WELL-X-TROL as system pressure passes the minimum pressure precharge. Only usable water is stored.

3.
When the pressure in the chamber reaches maximum system pressure, the pump stops. The WELL-X-TROL is filled to maximum capacity.

4.
When water is demanded, pressure in the air chamber forces water into the system. Since WELL-X-TROL does not waterlog and consistently delivers the maximum usable water, minimum pump starts are assured.

FIGURE 8.32 ▪ How diaphragm pressure tanks work. (*Courtesy of McGraw-Hill*)

Motor rating	Maximum starts per 24 hr. day	
	Single phase	Three phase
Up to ¾ hp	300	300
1 hp thru 5 hp	100	300
7½ hp thru 30 hp	50	100
40 hp and over		100

FIGURE 8.33 ▪ Recommended maximum number of times a pump should start in a 24-hour period. (*Courtesy of McGraw-Hill*)

| Pump discharge rate gpm (approx.) | Operating pressure—psig | | | | | |
| | 20/40 | | 30/50 | | 40/60 | |
	ESP I	ESP II	ESP I	ESP II	ESP I	ESP II
2.5	WX-104	WX-201	WX-104	WX-202	WX-104	WX-202
5	WX-201	WX-205	WX-202	WX-205	WX-202	WX-250
7	WX-202	WX-250	WX-203	WX-251	WX-205	WX-251
10	WX-203	WX-251	WX-205	WX-302	WX-250	WX-302
12	WX-205	WX-302	WX-250	WX-302	WX-251	WX-350
15	WX-250	WX-302	WX-251	WX-350	WX-251	WX-350
20	WX-251	WX-350	WX-302	(2)WX-251	WX-302	(2)WX-302
25	WX-302	(2)WX-302	WX-302	(2)WX-302	WX-350	(3)WX-251
30	WX-302	(2)WX-302	WX-350	(1)WX-302 (1)WX-350	WX-350	(2)WX-350
35	WX-350	(1)WX-302 (1)WX-350	WX-350	(2)WX-350	(2)WX-251	(3)WS-302
40	WX-350	(2)WX-350	(2)WX-251	(3)WX-302	(2)WX-302	(1)WX-302 (2)WX-350

FIGURE 8.34 ■ Sizing and selection information for perssure tanks. (*Courtesy of McGraw-Hill*)

WELL-X-TROL QUICK SIZING FORM
(We suggest you make an office copy of this page when ready to calculate.)

For selecting WELL-X-TROLs for a different running time than ESP I or ESP II, and/or at pressure ranges the same or different than 20/40, 30/50, 40/60:

THINGS YOU MUST KNOW

1. System flow rate (pump capacity or discharge) _____ GPM

2. Desired running time, in minutes and fractions of minutes (1.5 min. = 1 min. 30 sec.) _____ Min.

3. Pump cut-in, in gauge pressure _____ Psig

4. Pump cut-out, in gauge pressure _____ Psig

CALCULATING TANK SIZE

5. Multiply Line 1 by Line 2 and enter ESP Volume _____ ESP Vol.

6. Refer to Table 1. Find pressure factor for Line 3 and Line 4 and enter _____ P.F.

7. Divide Line 5 by Line 6 and enter minimum total WELL-X-TROL Volume _____ Gals.

8. Refer to Table 2 and select WELL-X-TROL model that is greater than Line 7 for "Total Volume" and Line 5 is less than "Maximum ESP Volume" _____ WX No.

9. Select precharge pressure _____ Psig

> NOTE: The precharge pressure *must* be adjusted to the pump cut-in pressure. For the following example reduce from 40 to 25 psig.

> *EXAMPLE:* A system flow will be delivered by a pump at a rate of 12.5 GPM. The pump switch is to be installed at the WELL-X-TROLL and has been determined to cut-in the pump at 25 psig. Its differential, or operating range, is 20 psi. It is desired to have the pump run *at least* one minute and 30 seconds every time it starts. Which WELL-X-TROL will provide "ESP"?

THINGS YOU MUST KNOW

1. System flow rate (pump delivery) 12.5 GPM

2. Desired running time, in minutes and fractions of minutes (1.5 min. = 1 min. 30 sec.) 1.5 Min.

3. Pump cut-in, in gauge pressure 25 Psig

4. Pump cut-out, in gauge pressure 45 Psig

CALCULATING TANK SIZE

5. Multiply Line 1 by Line 2 and enter ESP Volume 18.8 ESP Vol.

6. Find Pressure factor for Line 3 and Line 4 in Table 1, and enter 34 P.F.

7. Divide Line 5 by Line 6 and enter minimum total WELL-X-TROL volume 55.2 Gals.

8. Refer to Table 2 and select WELL-X-TROL model that is greater than Line 7 for "Total Volume" and Line 5 is less than "Maximum ESP Volume" WX-251

A WX-251 has a total volume of 62 gallons and a maximum ESP volume of 34 gallons.

FIGURE 8.35 ▪ Pressure tank sizing form. (*Courtesy of McGraw-Hill*)

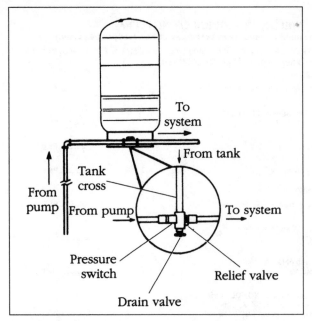

FIGURE 8.36 ■ Tank tee being used with a stand-type pressure tank. (*Courtesy of McGraw-Hill*)

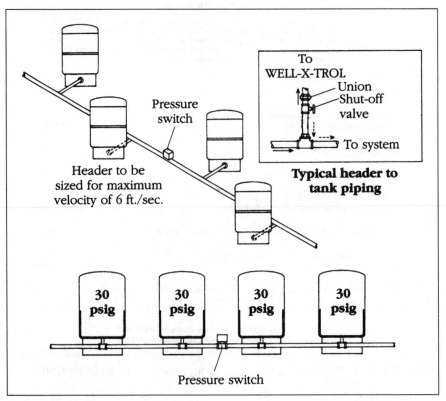

FIGURE 8.37 ■ Diagram of multiple pressure tanks being installed together. (*Courtesy of McGraw-Hill*)

chapter 9

CALCULATING MINIMUM PLUMBING FACILITIES

Calculating minimum plumbing facilities is a common part of a master plumber's job. Knowing and understanding what is required in a building is not only a requirement for plumbers. Architects and engineers are often the people who determine the requirements for a new building. Local plumbing codes dictate minimum plumbing facilities. All plumbers have to do is understand the information provided for them in their codebooks. The information given by the codes is fairly simple, but gaining a complete understanding of it can be a bit intimidating. If the process is approached too lightly, misconceptions can cause mistakes. The people responsible for determining what plumbing will be included in a building cannot afford to make mistakes.

It is common for plumbers to be provided with detailed blueprints when bidding jobs. The drawings will normally be submitted to a code enforcement office for approval. During this process, there are many ways for mistakes to be caught. If the person drawing the plans makes an error, the code officer who is working with the drawings is likely to find the problem. Plumbers bidding the job might catch the discrepancy.

Some jobs are not engineered. There are times when plumbers are expected to calculate the minimum plumbing needs for a building. Plus, plumbers who wish to

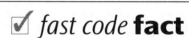

☑ *fast code* **fact**

When submitting plans and specifications with a permit application, it is the responsibility of the master plumber applying for the permit to make sure that the drawings, specifications, and details are accurate.

▶ *sensible* **shortcut**

Knowing the minimum plumbing requirements for a building is as simple as checking the tables in your local plumbing codebook. The process is not a complex one. Codes change and each jurisdiction can modify a code that is adopted, so check your current, local codebook for accurate information that applies to your specific area.

gain a master's license will have to pass an examination that is likely to require them to compute plumbing requirements. With this in mind, let's look at some tables that might be used to figure the requirements for plumbing fixtures in various types of buildings.

COMMERCIAL BUILDINGS OF MULTIPLE TENANTS

Commercial buildings of multiple tenants is our first topic of conversation (Fig. 9.1). This type of building can include a number of uses. Look at the table in Figure 9.1. You can see headings for water closets, lavatories, drinking fountains, and bathing fixtures. At first glance, the table seems simple enough, and it is not too difficult. But it can be confusing, so let's go through some sizing examples.

I want you to assume that there will be 62 people rated for the building that we are sizing. How many fixtures of each type will the building require? Take a moment to work the numbers, and then read the following results to see if you arrive at the same number that I do.

If you look under the heading for water closets, you will see that you need three for men and four for women. Also note the number 3 next to the water closet heading. Refer to Figure 9.2 for an explanation of the number. If you look at the number 3 in Figure 9.2, you will see conditions for various types of buildings within the general group that we are working with. For example, the statement requires urinals in male restrooms of restaurants, clubs, lounges, and so forth.

How many lavatories are needed in the restroom for women? The correct answer is three. Two lavatories are needed in male restrooms. How many bathing units are required? None, but our building will need a drinking fountain. Also note that drinking fountains are required on each floor, so this might increase the number of fixtures needed, depending upon building design. Pay attention to all details and footnotes when you use code charts and tables for sizing.

You probably already have a handle on this type of building, but let's do one more quick exercise. Using the same type of building, change the occupancy number to 125 people. What are the fixture requirements? We need four toilets in the male restroom and five in the female restroom. Two lavatories are required in the male restroom, and three are needed for the ladies. Drinking fountains are needed in the building. A minimum of two fountains is required.

☞ been there done that

As a young plumber, I thought the codebook was easy to deal with. Once I started being held responsible for my own code decisions, I found that the presentation of the code was not as clear as I once thought it was. Take some time to work with your codebook before you need it. Learn how to use the information in the code to your best advantage. This is best done with practice. Set yourself up with hypothetical circumstances and use your codebook to solve problems and answer questions. Check with a master plumber, when needed, to see if your solutions are correct. This will make your field work much easier as you come to rely on your code skills.

Building or occupancy[2]	Occupant content[2]	Water closets[3]			Lavatories[4]			Bathtubs, showers and miscellaneous fixtures	
		Persons (total)	Male	Female	Persons (total)	Male	Female	Drinking fountains Persons	Fixtures
Common toilet facilities for areas of commercial buildings of multiple tenants[8,9]	Use the sq ft per person ratio applicable to the single type occupancy(s) occupying the greatest aggregate floor area (Consider separately each floor area of a divided floor)	1–50	2	2	1–15	1	1	1–100	1
		51–100	3	4	16–35	1	2	101–250	2
		101–150	4	5	36–60	2	2	251–500	3
		For each additional 100 persons over 150, add	1	1.5^7	61–125	2	3	501–1000	4
					For each additional 120 persons over 125, add	1	1.5^7	Not less than one fixture each floor subject to access.	

FIGURE 9.1 ■ Minimum fixtures for commercial multi-tenant buildings. (*Courtesy of Standard Plumbing Code*)

Notes:

1. The figures shown are based upon one fixture being the minimum required for the number of persons indicated or any fraction thereof.
2. The occupant content and the number of required facilities for occupancies other than listed shall be determined by the plumbing official. Plumbing facilities in the occupancies or tenancies of similar use may be determined by the plumbing official from this table.
3. Urinals shall be required in male restrooms of elementary or secondary schools, restaurants, clubs, lounges, waiting room of transportation terminals, auditoriums, theaters, and churches at a rate equal to ½ of the required water closets in Table 407. Required urinals can be substituted for up to ½ of the required water closets. The installation of urinals shall be optional in the female restrooms of previously stated occupancies and shall be optional in both male and female restrooms of all other occupancies. Optional urinals may be substituted for up to ½ of the required water closets in the male and female restrooms.
4. Twenty-four linear inches (610 mm) of wash sink or 18 inches (457 mm) of a circular basin, when provided with water outlets for such space, shall be considered equivalent to 1 lavatory.
5. When central washing facilities are provided in lieu of washing machine connections in each living unit, central facilities shall be located for the building served at the ratio of not less than one washing machine for each 12 living units, but in no case less than two machines for each building of 15 living units or less. See 409.4.5.
6. A single facility consisting of one water closet and one lavatory may be used by both males and females in the following occupancies subject to the building area limitations:

Occupancy	Maximum building area (sq ft)
Office	1200
Retail Store (excluding service stations)	1500
Restaurant	500
Laundries (Self-Service)	1400
Beauty and Barber Shops	900

7. After totaling fixtures, round up any fraction to the next highest whole number of fixtures.
8. Common toilet facilities (separate for males and females) for each floor are acceptable in lieu of separate facilities required by this section only when the applicable building occupant has common access from within the building. When tenancies, rental units, etc., are to be provided with separate facilities of a partial nature, such facilities are not deductible from the total common facilities required.
9. (a) Applicable to small stand-up restaurants and similar occupancies.
   (b) Not applicable to do-it-yourself laundries, beauty shops and similar occupancies where persons must remain to receive personal services.
10. (a) Light manufacturing is applicable to those manufacturers manufacturing finished products which require no special equipment to handle single finished products may require special equipment to handle the products when packaged in containers containing multiple products.
    (b) Heavy manufacturing is applicable to those manufacturing processes requiring overhead cranes or similar equipment for the movement of raw materials and/or the finished products.

FIGURE 9.2 ■ Minimum fixtures requirement rules. (*Courtesy of Standard Plumbing Code*)

11. (a) Light Storage: Light storage is the storage of items which can be handled without the aid of special handling equipment such as cranes, forklifts or similar equipment.
    (b) Heavy Storage: Heavy storage is the storage of items which require special equipment for handling such as cranes, forklifts or similar equipment.
12. For other than industrial areas of the occupancy, see other applicable type occupancies (applicable to facilities provided due to inaccessibility of those in main or initial occupancy).
13. As required by the American Standard Safety Code for Industrial Sanitation in Manufacturing Establishments (ANSI Z4.1).
14. Where there is exposure to skin contamination with poisonous, infectious, or irritating materials, provide 1 lavatory for each 15 persons.
15. Laundry trays, 1 for each 50 persons. Slop sinks, 1 for each 100 persons.
16. For exclusively male or female dorms, the fixtures shall be double the amount required for the particular gender in a co-ed dorm.
17. If alcoholic beverages are to be served, facilities shall be as required for clubs or lounges.

FIGURE 9.2 ■ *(Continued)*

RETAIL STORES

The minimum fixture requirements for retail stores differ from the examples that we have just been working with. However, the concept and approach of computing the needs is same. Refer to Figure 9.3 for listings that pertain to retail stores. You can see that the table is very similar, in layout, to the one we have just been using. Pay particular attention to the number 6 at the heading of retail stores. Refer back to Figure 9.2 for an explanation of the note. You will find that one bathroom facility can be used by both males and females in certain types of occupancies. For example, an office with 1200 square feet, or less, can be served by a single restroom for both sexes. A retail store with 1500 square feet, or less, can also be served by one restroom, unless the store is classified as a service station. Other types of buildings that may qualify for a single bathroom are restaurants, self-service laundries, beauty salons, and barber shops. In all cases, the use of a single restroom is contingent on the square footage of the building. With this said, let's run through a sample sizing example.

Assume that our sample building will accommodate 59 people. Use the table in Figure 9.3 to determine the minimum number of plumbing fixtures required. For the purposes of this exercise, assume that the single-bathroom rule is not applicable. Go ahead, run the numbers, and then compare them with mine.

You should have found that the male restroom requires two water closets. A total of three water closets is needed for the female restroom. The male restroom is required to have only one lavatory, but the female restroom is required to have three lavatories. Only one drinking fountain is needed, subject to building design. By this, I mean that a drinking fountain is required on each

Retail stores[6]	200 sq ft per person						Drinking fountains	
	Persons (total)	Male	Female	Persons (total)	Male	Female	Persons	Fixtures
	1–35	1	1	1–15	1	1	1–100	1
	36–55	1	2	16–35	1	2	101–250	2
	56–80	2	3	36–60	1	3	251–500	3
	81–100	2	4	61–125	2	4	501–1000	4
	101–150	2	5	For each additional 200 persons over 125, add	1	1.75[7]	Not less than one fixture each floor subject to access.	
	For each additional 200 persons over 150, add	1	1.75[7]					

FIGURE 9.3 ■ Minimum fixtures for retail stores. (*Courtesy of Standard Plumbing Code*)

floor of the building, subject to access. If you did not arrive at these numbers, go over the table again and see if you can find the error in your calculations.

RESTAURANTS

There are a lot of restaurants in society. This is a common type of building for plumbers to work with. Finding the number of water closets and lavatories required in a restaurant is no more difficult than the other examples that we've been working with. However, there are additional requirements for restaurants. Essentially, you must check with your local code office and comply with minimum requirements that are established by the Board of Health.

Before we do a sizing example for a restaurant, let's discuss two alternative options. You will notice if you look at the headings in Figure 9.4, that the number 6 and the number 17 are next to the heading for restaurants. We've already discussed the option of number 6; it is the one where one bathroom might be allowed for use by both sexes. The option pertaining to number 17 are that if alcoholic beverages will be served, the establishment must meet facility requirements as set forth for clubs or lounges.

> ☑ *fast code* **fact**
>
> If you are a residential plumber, be aware that code requirements for commercial work can be very different from what you are accustomed to working with. Be sure that you are using the proper section of the code for the type of work that you are doing.

Now, let's do a sizing example. Assume that the restaurant we are working with is rated for 250 people. How many toilets are needed? Four water closets are required in each restroom. What is the required number of lavatories? The building calls for three lavatories in each restroom. Do you notice a difference in the ratings for restaurants, compared to the other types of buildings we have done thus far? If you review the tables, you will see that restaurants required, in almost all cases, the same number of fixtures for males as for females. In previous examples, female restrooms required more fixtures. This is not a big issue, just something I wanted to point out.

HOUSES

We've been dealing with commercial-type space, but let's switch over to houses. There are times when plumbers are the ones who must figure the minimum requirements for houses. This is especially true in very rural areas. Figure 9.5 provides the information needed to compute the fixture requirements for a typical, single-family home. The same table can be used to figure the fixture requirements for an apartment building. Check back to Figure 9.2 for explanations of the numbers noted in the headings of the table. There is no big secret to this table. Each home is required to have a minimum of one toilet, one lavatory, one bathing unit, one kitchen sink, and one connection for a washing machine.

If you look at the table closely, you will see that the basic minimums are required for each dwelling or dwelling unit. This means that each apartment

Restaurants[6.17] — 40 sq ft per person

Persons (total)	Male	Female	Persons (total)	Male	Female	Comply with board of health requirements
1–50	1	1	1–150	1	1	
51–100	2	2	151–200	2	2	
101–200	3	3	201–400	3	3	
201–300	4	4	For each additional 200 persons over 400 add	1	1	
For each additional 200 persons over 300 add	1	2				

FIGURE 9.4 ■ Minimum fixtures for restaurants. (*Courtesy of Standard Plumbing Code*)

Building or occupancy[2]	Occupant content[2]	Water closets[3]	Lavatories[4]	Bathtubs, showers and miscellaneous fixtures
Dwelling or Apt. House	Not Applicable	1 for each dwelling or dwelling unit	1 for each dwelling or dwelling unit	Washing machine connection per unit[5]. Bathtub or shower—one per dwelling or dwelling unit. Kitchen sink-one per dwelling or dwelling unit.

FIGURE 9.5 ■ Minimum fixtures for homes and apartments. (*Courtesy of Standard Plumbing Code*)

in a building must be equipped with the same minimum requirements that would be found in a home. Of course, local codes may offer a different ruling, so always check your local code requirements before designing or installing plumbing systems.

DAY-CARE CENTERS

Day-care centers, pre-schools, and nursery schools all fall under the same classification when computing minimum needs for plumbing fixtures. Figure 9.6 shows you the formulas for figuring the number of fixtures needed. It's a simple table. Basically, you supply one toilet and one lavatory for each 15 occupants of the building. If the school will have 30 occupants, you must install two toilets and two lavatories. When 45 people will be in the school, you need three toilets and three lavatories. This is one of the easiest sizing exercises going.

ELEMENTARY AND SECONDARY SCHOOLS

Requirements for elementary and secondary schools are a bit more complex than those applying to pre-schools. Even so, the process of sizing the fixture needs is not difficult. Look at the table in Figure 9.7. Notice that the table is very similar to the ones we have been using. An equal number of toilets and lavatories is required in both male and female restrooms. One drinking fountain is required for every three classrooms in the school. It is also a requirement that a drinking fountain be located on each floor of the building. All you have to do in order to figure fixtures for a school is look at the number of occupants and reference it next to the number of fixtures. For example, if you have 98 occupants, you need three toilets and two lavatories in each bathroom.

OFFICES AND PUBLIC BUILDINGS

Offices and public buildings may be allowed to have only one bathroom, subject to the size and use of the building. Refer to Figure 9.2, line number 6, for a complete description of possible options in using a single bathroom. The table in Figure 9.8 shows you sizing information for offices and public buildings where multiple bathrooms are used. Feel up to another sizing example? Well, let's try a couple with the table in Figure 9.8.

Schools: Pre-School, Day Care or Nursery	Average Daily Attendance	Each 15 children or fraction thereof	1 Fixture	Each 15 children or fraction thereof	1 Fixture

FIGURE 9.6 ■ Minimum fixtures for pre-schools. (*Courtesy of Standard Plumbing Code*)

Schools: Elementary & Secondary	Average Daily Attendance							One drinking fountain for each 3 classrooms, but not less than one each floor
		Persons (total)	Male	Female	Persons (total)	Male	Female	
		1–50	2	2	1–120	1	1	
		51–100	3	3	121–240	2	2	
		101–150	4	4	For each additional 120 persons over 240, add	1	1	
		151–200	5	5				
		For each additional 50 persons over 200, add	1	1				

FIGURE 9.7 ■ Minimum fixtures for elementary and secondary schools. (*Courtesy of Standard Plumbing Code*)

Office[6] and public buildings	100 sq ft per person	Persons (total)	Male	Female	Persons (total)	Male	Female	Drinking fountains	
								Persons	Fixtures
		1–15	1	1	1–15	1	1	1–100	1
		16–35	1	2	16–35	1	2	101–250	2
		36–55	2	2	36–60	2	2	251–500	3
		56–100	2	3	61–125	2	3	Not less than one fixture each floor subject to access.	
		101–150	3	4	For each additional 120 persons over 125 add	1	1.5^7		
		For each additional 100 persons over 150 add	1	1.5^7					

FIGURE 9.8 ■ Minimum fixtures for offices and public buildings. *(Courtesy of Standard Plumbing Code)*

Okay, assume that the public building we are working with will be rated for 75 people. What are the fixture requirements? We will need two toilets in the male restroom and three in the female restroom. How many lavatories are required? Two lavatories are needed in the male restroom and three are required in the female restroom. One drinking fountain is required, but others may be required if the building has more than one floor level, since a fountain is required on each floor of the building. That wasn't too hard, was it? Now let's try an example with a larger occupancy load.

In this example, assume that there will be 250 occupants. The amount of water closets needed in the male restroom is four. How many are need in the female restroom? Looks like 5½ toilets, right? Well, it is, but you have to round up to the next nearest whole number. In other words, you would need six toilets in the female restroom. What are the needs for lavatories? The male restroom requires four lavatories, and the female lavatory need is six. How many drinking fountains are needed in a building that has only a single floor? Three fountains are required.

CLUBS AND LOUNGES

When you are dealing with clubs and lounges, you must pay attention to Board of Health requirements. Remember that if a restaurant will be serving alcoholic beverages, the building will be treated as a club or lounge for fixture requirements. I should also point out that the tables we are using are from the Standard Plumbing Code. Local codes vary, so don't use these tables for your actual work. I'm providing the tables for the sake of examples, not as the final word.

I'm not going to continue doing routine examples of table use. You should understand the basic concepts now. However, I will touch on the remaining categories and provide you with sample tables for determining minimum plumbing fixtures. Figure 9.9 is a table set up for clubs and lounges. There is nothing unusual about the table, so apply the same principles that we have been working with.

LAUNDRIES

Do-it-yourself laundries are required to have at least one drinking fountain and one service sink. Figure 9.10 will give you the basics for sizing fixture requirements of do-it-yourself laundries. Notice that this type of laundry might be allowed to operate with a single bathroom.

HAIR SHOPS

Hair shops, like beauty salons and barber shops are required to have a drinking fountain and a service or utility sink. Figure 9.11 shows the basic requirements for fixtures in these types of buildings. It is worth noting that only one lavatory is required for each bathroom, regardless of the occupancy load. It is also possible that beauty shops and barber shops might be required to maintain only one restroom for occupants.

Building or occupancy[2]	Occupant content[2]	Water closets[3]			Lavatories[4]			Bathtubs, showers and miscellaneous fixtures
		Persons (total)	Male	Female	Persons (total)	Male	Female	Comply with board of health requirements
Clubs, lounges, and restaurants with club or lounge[6]	40 sq ft per person	1–50	2	2	1–150	1	1	
		51–100	3	3	151–200	2	2	
		101–300	4	4	201–400	3	3	
		For each additional 200 persons over 300, add	1	2	For each additional 200 persons over 400, add	1	1	

FIGURE 9.9 ■ Minimum fixtures for clubs and lounges. (*Courtesy of Standard Plumbing Code*)

Do-it-yourself laundries[6]	50 sq ft per person	Persons (total)	Male	Female	Persons (total)	Male	Female	One drinking fountain and one service sink.
		1–50	1	1	1–100	1	1	
		51–100	1	2	101–200	2	2	

FIGURE 9.10 ■ Minimum fixtures for do-it-yourself laundries. (*Courtesy of Standard Plumbing Code*)

Beauty shops and barber shops[6] 50 sq ft per person	Persons (total)	Male	Female	Persons (total)	Male	Female	One drinking fountain and one service or other utility sink.
	1–35	1	1	1–75	1	1	
	36–75	2	2				

FIGURE 9.11 ■ Minimum fixtures for hair-care establishments. *(Courtesy of Standard Plumbing Code)*

WAREHOUSES, FOUNDRIES, AND SUCH

Warehouses, foundries, manufacturing buildings, and similar buildings have some special requirements. For example, a shower must be provided for each 15 people who may be exposed to excessive heat or to skin contamination with poisonous, infectious, or irritating material. When you look at the table in Figure 9.12 you will see a number of numbers at the topic headings. Refer back to Figure 9.2 for an understanding of the special notes. An example of such a note is number 14 in the list of Figure 9.2. It says that one lavatory must be supplied for every 15 people who may have exposure to skin contamination with poisonous, infectious, or irritating materials. Refer all of the special notes before you begin figuring your fixture needs.

LIGHT MANUFACTURING

Buildings used for light manufacturing are affected by the special notes listed in Figure 9.2. Pay attention to all the note references in the table labeled as Figure 9.13. You will see that the sizing table is like the others that we have been using and is just as easy to negotiate.

DORMITORIES

Dormitories require the use of one laundry tray for each 50 people and one slop sink for each 100 people. However, washing machines can be used in lieu of laundry trays. This information is found in Figure 9.14 and Figure 9.2. It is also required that dormitories which are for the exclusive use of one sex or the other shall have double the number of fixtures listed under the gender-specific restrooms in the table. There are also rulings in Figure 9.14 pertaining to bathtubs and showers. You will find that sizing dormitories is not difficult, but that it does involve some rules that we have not previously used.

GATHERING PLACES

Gathering places, such as churches, theaters, auditoriums, and similar places can be sized for plumbing fixtures by using the information in Figure 9.15. This reference table is straightforward and holds no surprises. If at any time you are not sure how to title a building's classification, check with your local code enforcement office. Remember also to verify local standards for sizing requirements. Given the proper information from your local code, you should have no trouble determining the minimum fixture requirements for buildings.

Heavy manufacturing,[10] warehouses,[11] foundries, and similar establishments[12,14]	Occupant content per shift, substantiated by owner. Also see 407.3.2	Persons (total)	Male	Female	Persons (total)	Male	Female	One drinking fountain for each 75 persons. One shower for each 15 persons exposed to excessive heat or to skin contamination with poisonous, infectious, or irritating material.
		1–10	1	1	1–15	1	1	
		11–25	2	1	16–35	2	1	
		26–50	3	1	36–60	3	1	
		51–75	4	1	61–90	4	1	
		76–100	5	1	91–125	5	1	
		For each additional 60 persons over 100, add	1	0.1[7]	For each additional 100 persons over 125 add	1	0.1[7]	

FIGURE 9.12 ■ Minimum fixtures for heavy manufacturing. (*Courtesy of Standard Plumbing Code*)

Building or occupancy[2]	Occupant content[2]	Water closets[3]			Lavatories[4]			Bathtubs, showers and miscellaneous fixtures
		Persons (total)	Male	Female	Persons (total)	Male	Female	
Light mfg.[10] Light Warehousing[11] and workshops, etc.[12,13]	Occupant content per shift, substantiated by owner Also see 407.3.2	1–25	1	1	1–35	1	1	One drinking fountain for each 75 persons. One shower for each 15 persons exposed to excessive heat or to skin contamination with poisonous, infectious, or irritating material.
		26–75	2	2	36–100	2	2	
		76–100	3	3	101–200	3	3	
		For each additional 60 persons over 100 persons add	1	1	For each additional 100 persons over 200 persons add	1	1	

FIGURE 9.13 ▪ Minimum fixtures for light manufacturing. (*Courtesy of Standard Plumbing Code*)

Dormitories[15]	Persons (total)	Male[16]	Female[16]	Persons (total)	Male	Female	Washing machines may be used in lieu of laundry tubs.[15]
50 sq ft per person (calculated on sleeping area only)	1–10	1	1	1–12	1	1	One shower for each 8 persons. In women's dorms add tubs in the ratio 1 for each 30 females. Over 150 persons add 1 shower for each 20 persons.
	11–30	1	2	13–30	2	2	
	31–100	3	4	For each additional 30 persons over 30 add			
	For each additional 50 persons over 100, add	1	1		1	1	

FIGURE 9.14 ■ Minimum fixtures for dormitories. (*Courtesy of Standard Plumbing Code*)

	Persons (total)	Male	Female	Persons (total)	Male	Female	Drinking fountains Persons	Fixtures
Theaters, auditoriums, churches, waiting rooms at transportation terminals and stations	70 sq ft per person (calculated from assembly area). Other areas considered separately. (See Office or Public Buildings.)							
	1–50	2	2	1–200	1	1	1–100	1
	51–100	3	3	201–400	2	2	101–350	2
	101–200	4	4	401–750	3	3	Over 350 add one fixture for each 400.	
	201–400	5	5	Over 750 persons lavatories shall be required at a number equal to not less than ⅓ of total of required water closets and urinals.				
	For each additional 250 persons over 400, add	1	1					

FIGURE 9.15 ▪ Minimum fixtures for gathering places. (*Courtesy of Standard Plumbing Code*)

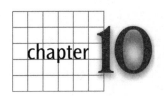

chapter 10

CALCULATING PROPER FIXTURE
SPACING AND PLACEMENT

Standard fixture layouts are dictated by local plumbing codes. Plumbing codes require certain amounts of space to be provided in front of and beside plumbing fixtures. The rules for standard fixtures are different than those used to control the installation of handicap fixtures. We will use this chapter to cover the essentials of standard fixtures and address the topic of handicap fixtures in the next chapter. For now, just concentrate on typical fixture installations when you review the information in this chapter. Before we get into deep details, I want to remind you to consult your local plumbing code for requirements specific to your region. The numbers I give you here are based on code requirements, but they may not be from the code that is enforced in your area.

If you work mostly with new construction, you probably work from blueprints. When this is the case, fixture locations are usually indicated and approved before a job is started. But, remodeling jobs can require plumbers to make on-site determinations for fixture placement. A contractor might ask you to provide spacing requirements for small jobs. Knowing how to do this is important. For example, if a builder showed you a sketch, like the one in Figure 10.1, would you be able to assign numbers to the areas around the fixtures? How wide would the compartment where the toilet is housed be required to be? The answer is 30 inches. This is common knowledge for many plumbers, and codebooks define the distance. So, even if you don't know the spacing requirements off the top of your head, you can always consult your local code for the answers.

A general rule for toilets is that there must be at least 15 inches of clear space on either side of the center of the drain for the toilet. This equates to a total space of 30 inches (Fig. 10.2). Now, how much clearance is needed in front of a toilet? The normal answer is 18 inches (Fig. 10.3). Some bathrooms are small. This can create a problem for plumbers, especially if you are remodeling the bathroom with new fixtures or possibly different types of

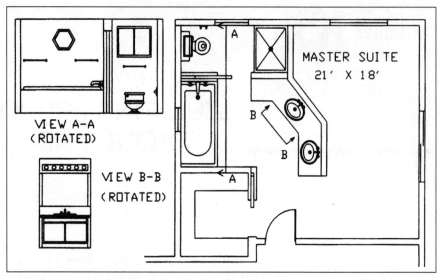

FIGURE 10.1 ■ A typical bathroom layout. (*Courtesy of McGraw-Hill*)

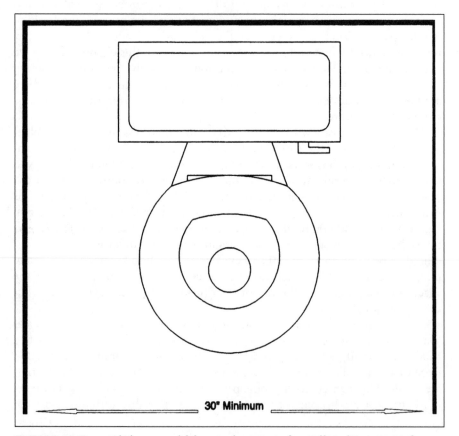

FIGURE 10.2 ■ Minimum width requirements for toilet. (*Courtesy of McGraw-Hill*)

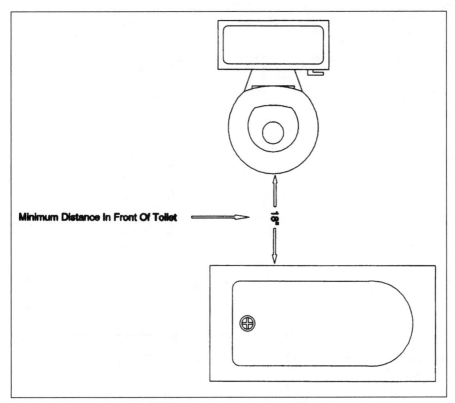

Minimum Distance In Front Of Toilet

18"

FIGURE 10.3 ▪ Minimum distance in front of toilet. (*Courtesy of McGraw-Hill*)

fixtures. Getting your rough-in for the fixtures right is crucial to the job. If you install a drain for a toilet and find out when you go to set fixtures that there is inadequate space for the toilet to comply with the plumbing code, there could be a lot of work and expense required to correct the situation.

When you are laying out plumbing fixtures, you should concentrate on what you are doing. Get to know your code requirements and check the fixture placement in all directions. Figure 10.4 shows how a legal layout might look. In contrast, Figure 10.5 shows what would result in an illegal layout. Notice that the distance from the edge of the vanity is only 12 inches from the center of the toilet. To

▶ *sensible* **shortcut**

When installing a drain for a standard toilet, the center of the drain should be 12 inches from the wall that the toilet backs up to. It is possible to get toilets with different rough-in dimensions. For example, you could get a toilet where the distance from the back wall to the center of the drain would be 10 inches. Another option is a toilet with a rough-in of 14 inches.

meet code, the distance must be at least 15 inches. A problem like this might be avoided by using a smaller vanity. If the potential problem is caught on paper, before pipes are installed, it is much easier and less costly to correct.

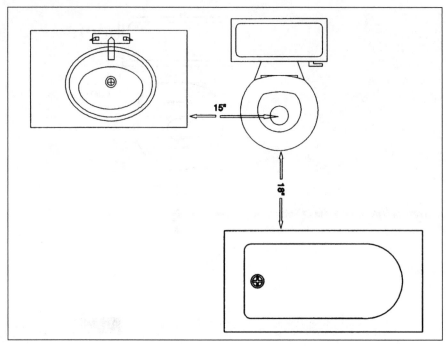

FIGURE 10.4 ■ Minimum distances for legal layout. (*Courtesy of McGraw-Hill*)

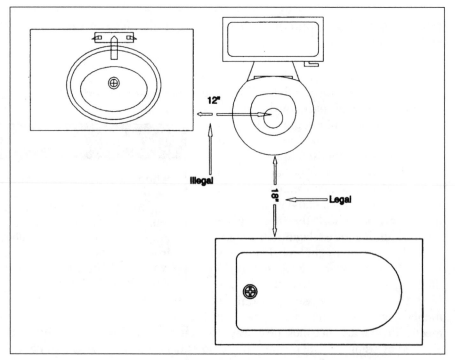

FIGURE 10.5 ■ Illegal fixture spacing. (*Courtesy of McGraw-Hill*)

Something else to consider when setting fixtures is their overall alignment. The plumbing codes not only require certain defined standards, they also deal with topics such as workmanship. This means that a job could be rejected if the fixtures are installed in a sloppy manner. Figure 10.6 shows a toilet where the flush tank is not installed with equal distance from the back wall. A proper installation would have the toilet tank set evenly, with equal distance from the back wall, as is indicated in Figure 10.7.

☞ been there **done that**

There are times when space is at a premium. Consider using corner fixtures, such as a corner shower or corner toilet. This can buy you enough space to make a remodeling job work.

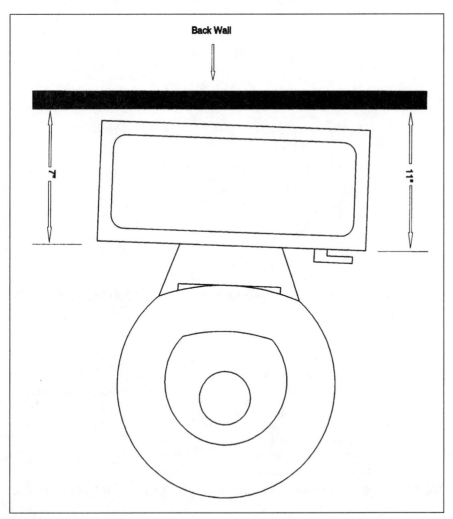

FIGURE 10.6 ■ Improper toilet alignment. (*Courtesy of McGraw-Hill*)

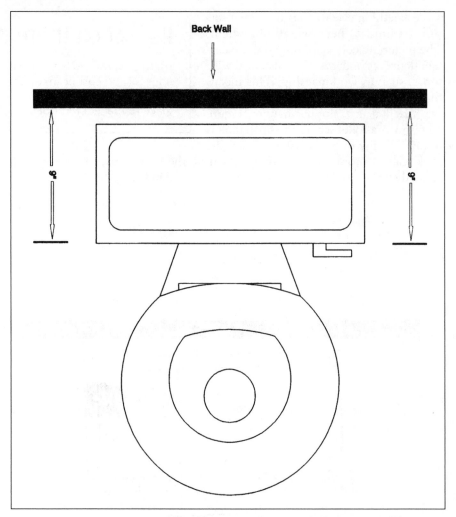

FIGURE 10.7A ■ Proper toilet alignment. (*Courtesy of McGraw-Hill*)

Measurement	Minimum distance (in inches)
From center of drain to any object on either side	15
From front of fixture to any object in front of it	18
Width of a privacy compartment	30
Depth of a privacy compartment	60

FIGURE 10.7B ■ Clearances for water closets. (*Courtesy of McGraw-Hill*)

CLEARANCES RELATED TO WATER CLOSETS

Let's talk about clearances related to water closets. There's not a lot to go over, so this can move along quickly. Remember that we are talking about standard plumbing fixtures here, not handicap fixtures. The minimum distance required from the cen-

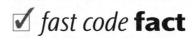

✓ *fast code* **fact**

Workmanship is an element of the plumbing code that some workers fail to consider. A code officer can fail a job for its inspection if the workmanship is sloppy, so do neat work.

ter of a toilet drain to any obstruction on either side is 15 inches. Measuring from the front edge of a toilet to the nearest obstruction must prove a minimum of 18 inches of clear space. When toilets are installed in privacy stalls, you must make sure that the compartments are at least 30 inches wide and at least 60 inches deep. That's all there is to a typical toilet layout (Fig. 10.7).

URINALS

Urinals must have a minimum distance of 15 inches from the center of the drain to the nearest obstruction on either side. If multiple urinals are mounted side by side, there must be a minimum of 30 inches between the two urinal drains. The required clearance in front of a urinal is 18 inches.

LAVATORIES

Lavatories are not affected by side measurements, unless other types of plumbing fixtures are involved. The minimum distance in front of a lavatory should not be less than 18 inches. Obviously, minimum requirements are just that, minimums. It is best when more space can be dedicated to a bathroom in order to make the fixtures more user-friendly.

> *sensible* **shortcut**

A pedestal lavatory can be a good option if you have limited space to work with. Many pedestal lavatories are available in sizes that are small enough to give you the extra inch, or two, that you may need.

KEEPING THE NUMBERS STRAIGHT

Keeping the numbers straight for standard plumbing fixtures doesn't require a lot of brain space. There are very few numbers to commit to memory. The process gets somewhat more complicated when you are dealing with handicap or "accessible" fixtures. We're done with standard fixtures, so let's explore accessible fixtures.

HANDICAP FIXTURE LAYOUT

Layouts for handicap plumbing fixtures require more space than what would be needed for standard plumbing fixtures. When you are planning the installation of accessible fixtures you must take many factors into consideration.

There are regulations pertaining to door widths, compartment sizes, fixture locations, and so forth. These rules and regulations are generally provided in local plumbing codes. It's not necessary for you to commit all the measurements to memory, but you need to be aware of them and know where to find the figures when they are needed for design issues.

When you are dealing with a building that requires the installation of handicap plumbing fixtures, you find yourself spread between the local building code and the local plumbing code. The two codes overlap when it comes to handicap facilities. Are you, as a plumber, responsible for the carpentry work? It depends upon how you look at it. You probably have no responsibility for the width of a door used for access to a bathroom where handicap fixtures will be installed. But, the width of a toilet compartment will affect you and your work. Most trades work well together on most jobs, but this is not always the case. If you know that rough framing done by a carpentry crew is going to prohibit you from installing fixtures with proper placement, you should talk to someone about the impending problem. Whether you talk to the carpenters, a carpentry foreman, a job superintendent, or a general contractor, you should raise the question of what you perceive to be a problem with the framing work. The quicker potential problems can be caught, the easier it will be to correct them.

> ▶ *sensible* **shortcut**

If you will be installing grab bars as a part of your plumbing installation, you must remember to install suitable backing supports in the stud walls prior to wall coverings being applied to the studs. I recommend 2 x 6 or 2 x 8 lumber for this purpose. The larger the backing is, the easier it is to screw into once the walls are closed up. Make notes on where the backing is installed, so that you can find it when setting fixtures.

Who is responsible for the installation of grab bars? It could be the plumbers or the carpenters. This is an issue that must be addressed before a bid is given for a job. Grab bars are not inexpensive, so don't make a mistake by omitting them from a bid price where the person you are bidding the job for expects you to include the bars and their installation in your bid. There's something else to consider on this issue. If you are responsible for the grab bars, you are also responsible for installing proper supports in the framed walls during your rough-in work. Some type of backing, such as a length of framing lumber, must be installed in the wall cavity where a grab bar will be installed. Without the backing, the grab bars will not be solid. Finding out that there is no solid support to secure a grab bar to after a job has finished wall coverings is going to be a real problem. I've seen many jobs where backing wasn't installed for wall-hung lavatories and grab bars. This is an expensive and embarrassing mistake.

FACILITIES FOR HANDICAP TOILETS

Let's talk about the facilities for handicap toilets (Fig. 10.8). When a handicap toilet is installed in a privacy compartment, the minimum net clear opening for the compartment must be at least 32 inches wide. The door of the

4094 Atlas Elongated Rim
- 18" rim height handicapped.
- 12" rough-in.
- Anti-Siphon ballock.
- 3.5 G.P.F.

FIGURE 10.8 ■ Handicap toilet.
(*Courtesy of McGraw-Hill*)

compartment must swing out, away from the toilet. The width of such a compartment should be 36 inches, with a depth of 60 inches. Unlike a standard toilet where the side clearance is 15 inches, handicap toilets require a side distance of 18 inches.

Grab bars must be installed at a height of no less than 33 inches and no more than 36 inches above the finished floor. The bars must have a minimum length of 42 inches. They must be mounted on both sides of the compartment. When the bars are mounted, they must be mounted a maximum of 12 inches from the rear wall and extend a minimum of 54 inches from the rear wall. A rear grab bar, of at least 36 inches in length, must also be installed. This grab bar must be no more than 6 inches from the closest sidewall and extend a minimum of 24 inches beyond the centerline of the toilet away from the closest sidewall.

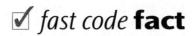

☑ *fast code* **fact**

Clearances for handicap fixtures differ from those used for conventional fixtures. Don't make the mistake of using the wrong spacing tables when selecting fixture locations.

Toilets approved for handicap installations must be higher than a normal toilet. Most of them are 18 inches tall, but the allowable range is anything between 16 and 20 inches above the finished floor. Rules for single-occupant arrangements vary a little from commercial installations. As always, check your local plumbing code for exact regulations in your region.

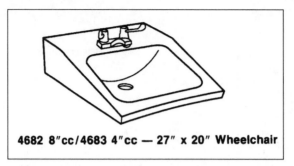

4682 8"cc/4683 4"cc — 27" x 20" Wheelchair

FIGURE 10.9 ■ Handicap lavatory. (*Courtesy of McGraw-Hill*)

LAVATORIES

Lavatories installed for handicap use must be of a type that is accessible by a person in a wheelchair (Fig. 10.9). The minimum clear space in front of a lavatory must be 30 inches by 30 inches. This is based on a measurement made from the front face of the lavatory, counter, or vanity. Measuring from the finished floor to the top edge of a lavatory or counter should result in a measurement of 35 inches. How much clearance is required under the lavatory? Unobstructed knee clearance with a minimum of 29 inches high by 8 inches deep should be provided. Toe clearance should be a minimum of 9 inches high by 9 inches deep, provided from the lavatory to the wall.

Additional requirements for a handicap lavatory require that all exposed hot-water piping be insulated. Faucets should be installed so that they are no more than 25 inches from the front face of the lavatory, counter, or vanity. And, the faucet must be able to be turned on and off with a maximum force of five pounds. Now, what happens if the lavatory is installed in a privacy compartment of a toilet? When a lavatory is installed in a compartment, the lavatory must be located against the back wall, adjacent to the water closet. The edge of the lavatory must have a minimum of 18 inches of clear space, measured from the center of the toilet.

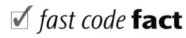 **fast code fact**

Faucets installed for handicap use must be approved for the use. This normally means either a single-handle faucet or a faucet with blade handles.

KITCHEN SINKS

Kitchen sinks require a minimum clear space in front of them that must be 30 inches by 30 inches. This is based on a measurement made from the front face of the kitchen sink, counter, or vanity. Measuring from the finished floor to the top edge of a kitchen sink or counter should result in a measurement of 34 inches, maximum. Unobstructed knee clearance with a minimum of 29 inches high by 8 inches deep should be provided. Toe clearance should be a minimum of 9 inches high by 9 inches deep, provided from the sink to the wall.

Additional requirements for a handicap kitchen sink require that all exposed hot-water piping be insulated. Faucets should be installed so that they are no more than 25 inches from the front face of the lavatory, counter, or vanity. And, the faucet must be able to be turned on and off with a maximum force of five pounds.

BATHING UNITS

Bathing units for handicap use are required to be equipped with grab bars. Bathtubs and showers for handicap use are often different in size and equipment from what you would find in a standard fixture (Fig. 10.10, Fig. 10.11). The minimum clear space in front of a bathing unit is 30 inches from the edge of the enclosure away from the unit and 48 inches wide. If a situation exists where a bathing unit is not accessible from the side, the clear space in front of the unit must be increased to a minimum of 48 inches. Faucets for showers and bathtubs must be equipped with a hand-held shower. The hose for these showers must be a minimum of 60 inches in length. The faucets must be able to be opened and closed with a maximum force of five pounds.

Grab bars are required in handicap bathing units. Diameters and widths of grab bars must be a minimum of 1.25 inches and a maximum of 1.5 inches. The bars must be spaced 1.5 inches from the wall. It is not allowable for the bars to rotate. All bars used must be approved for the intended use.

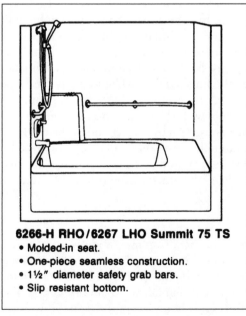

6266-H RHO / 6267 LHO Summit 75 TS
- Molded-in seat.
- One-piece seamless construction.
- 1½″ diameter safety grab bars.
- Slip resistant bottom.

FIGURE 10.10 ■ Handicap bathtub.
(Courtesy of McGraw-Hill)

6066-H RHO/6067 LHO Summit 36S
- One-piece seamless construction.
- Fold-down bench.
- 1½" diameter safety grab bars
- Meets ANSI standard A117.1-80.
- Slip resistant floor.

FIGURE 10.11 ■ Handicap shower.
(*Courtesy of McGraw-Hill*)

Bathtubs

Bathtubs for handicap use are required to have a seat. The seat may be built in or a detachable model. Grab bars with a minimum length of 24 inches must be mounted against the back wall, in line with each other and parallel to the floor. One of the bars, the top one, must be mounted a minimum of 33 inches and a maximum of 36 inches above the finished floor. The lower bar must be mounted 9 inches above the flood-level rim of the bathtub. A grab bar must be mounted at each end of the bathtub, with the bars being the same height as the top bar on the back wall. The bar used on the faucet end of the tub must be at least 24 inches long. A bar mounted at the other end of the tub must be at least 12 inches long. Faucets must be mounted below the grab bar. If a seat is installed at the end of a bathtub, the grab bar for that end must be omitted.

Showers

There are two basic types of showers for handicap use. Wide shower enclosures are one type, and square shower enclosures are the other. Shower stalls may be made on site or purchased as pre-fab units (Fig. 10.12). When a wide shower enclosure is used, it must have a minimum width of 60 inches. The depth must be no less than 30 inches. Thresholds are prohibited. Showers of this type must be made to allow wheelchairs to enter the enclosure. Shower valves must be mounted on the back wall. The minimum distance for the valve

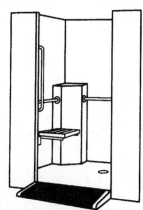

6950 RH Seat/6951 LH Seat Liberte
- Has fold-down seat. Placed at 18″ height for easy transfer from wheelchair to seat.
- Two built-in soap shelves.
- One vertical and three horizontal grab bars.
- Inside diameter of 5′ for easy wheelchair turn inside stall.
- Entry ramp 36″ wide with gentle 8.3% grade.
- Lipped door ledge to prevent rolling out of stall.
- Anti-skid floor mat included.
- White.
- Optional dome (6951) available.

FIGURE 10.12 ■ Handicap shower with seat and ramp. (*Courtesy of McGraw-Hill*)

from the shower floor is 38 inches, with a maximum height of 48 inches. A grab bar must be mounted along the entire length of the three walls that form the enclosure. All grab bars are to be set at least 33 inches above the shower floor, but not more than 36 inches above the floor. And, the bars shall be mounted parallel to the shower floor.

A shower enclosure that is square in design has to be at least 36 inches square. Seats for this type of shower may have a seat with a maximum width of 16 inches. The seat must be mounted along the entire length of the shower. Seat height is established as a minimum of 17 inches above the shower floor, with a maximum height of 19 inches. Grab bars must be installed to extend from the edge of the seat

> ▶ *sensible* **shortcut**
>
> Don't order handicap fixtures until you are sure that they are approved for use in your jurisdiction. When in doubt, check with your local code officer to confirm approval for specific fixtures.

around the sidewall opposite the seat. These bars must be at least 33 inches above the shower floor, and not more than 36 inches above the floor. A shower valve must be mounted on the sidewall opposite the seat. The minimum height of the shower valve shall be 38 inches above the floor. A maximum height of 48 inches is allowed for the installation of a shower valve.

DRINKING FOUNTAINS

Drinking fountains installed for handicap use shall be installed so that the spout is no more than 36 inches above the finished floor. The spout must be located in the front of the fountain. It is required that the flow of water from the spout shall rise at least 4 inches. Controls for operating the fountain may be mounted on front of the fountain or to the side, so long as the control is side-mounted near the front of the fountain. All handicap fountains require a minimum clear space of 30 inches in front of the fixture. The measurement is made from the front of the unit by 48 inches wide. If a fountain protrudes from a wall, the clear space may be reduced from a width of 48 inches to a width of 30 inches. Handicap fixtures require more attention than standard fixtures. Keeping all the clearances straight in your head can be confusing. Refer to your local codebook whenever you need clarification on a measurement.

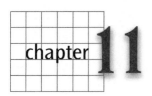

chapter 11

MATH FOR MATERIALS

When the door to pipe, tubing, and fittings is opened, there is a lot to learn. Some of the information is used on a frequent basis, and some of it turns up only in remote situations. We are going to open that door in this chapter. You are going to learn about various types of pipe and tubing. I expect that you will find some of the data fascinating and some of it boring. Use what you want. I will present the details in the most user-friendly manner that I can. Tables will be used to make the reference material fast and easy to see and understand. There's much to learn, so let's get started.

THE UNIFIED NUMBERING SYSTEM

Are you aware of the Unified Numbering System (UNS)? This is a system that is meant to correlate the many metal alloy numbering systems that are being used in our country. I could go into a long discussion on this, but I believe that a simple table will give you enough information for now. Figure 11.1 shows you the various categories of alloys. If you look to the left of the table, you will see letters. The letters are the beginning for understanding types of alloys. For example, if a rating starts with the letter C, it is referring to copper. Seeing a letter F at the beginning of a rating indicates cast-iron.

METRIC SIZES

Metric sizes are common in many places of the world. Plumbers in the United States still work primarily with customary measurements in terms of inches. However, you may find times when metric equivalents are useful. For this reason, I'm providing Figure 11.2

▶ sensible **shortcut**

If you are like me, you never took to metric measurements and don't really want to begin to now. This is okay. The sensible shortcut for this situation is conversion tables, which are abundant in this book. You don't have to do the math when you can consult a conversion table.

The first letter (followed by five digits)	Alloy category (assigned to date)
Axxxxx	Aluminum and its alloys
Cxxxxx	Copper and its alloys
Exxxxx	Rare-earth metals, and similar metals and alloys
Fxxxxx	Cast irons
Gxxxxx	AISI and SAE carbon and alloy steels
Hxxxxx	AISI and SAE H-steels
Jxxxxx	Cast steels (except tool steels)
Kxxxxx	Miscellaneous steels and ferrous alloys
Lxxxxx	Low-melting metals and their alloys
Mxxxxx	Miscellaneous nonferrous metals and their alloys
Nxxxxx	Nickel and its alloys
Pxxxxx	Precious metals and their alloys
Rxxxxx	Reactive and refractory metals and their alloys
Sxxxxx	Heat- and corrosion-resistant steels (including stainless), valve steels and iron-based "superalloys"
Txxxxx	Tool steels (wrought and cast)
Wxxxxx	Welding filler metals
Zxxxxx	Zinc and its alloys

FIGURE 11.1 ■ UNS metal family designations. (*Courtesy of McGraw-Hill*)

for your use in comparing common measurements from the United States to metric measurements.

THREADED RODS

Threaded rods are often used to hang various types of pipe. If the size of the threaded pipe is too small in diameter and in its ability to support a proper amount of weight, the use of the rod can be very destructive. If you have a need to choose threaded rod for hanging pipe, you should find the information in Figures 11.3 and 11.4 very helpful.

FIGURING THE WEIGHT OF A PIPE

Figuring the weight of a pipe and its contents is necessary when you are choosing the needed strength of a pipe hanger. There is a formula that you can use to accomplish this goal. Let's say that you want to know how much a

Nominal pipe size (NPS), in IP	ASHRAE std. wt. size, mm	AWWA pipe size, mm	NFPA pipe size, mm	ASTM copper tube size, mm	Nominal pipe size DN, mm
⅛	—	—	—	6	6
³⁄₁₆	—	—	—	8	8
¼	8	—	—	10	10
⅜	10	—	—	12	12
½	15	12.7 & 13	12	15	15
⅝	—	—	—	18	18
¾	20	—	—	22	20
1	25	25	25 & 25.4	28	25
1¼	32	—	33	35	32
1½	40	45	38 & 38.1	42	40
2	50	50 & 50.8	51	54	50
2½	65	63 & 63.5	63.5 & 64	67	65
3	80	75	76 & 80	79	80
3½	—	—	89	—	90
4	100	100	102	105	100
4½	—	114.3			115
5	—	—	127	130	125
6	150	150	152	156	150
8	200	200	203	206	200
10	250	250	—	257	250
12	300	300	305	308	300
14	—	350	—		350
18	—	400	—		400
18	—	—	—		450
20	—	500	—		500
24	—	600	—		600
28					700
30					750
32					800
36					900
40					1000
44					1100
48					1200
52					1300
56					1400
60					1500

FIGURE 11.2 ■ Equivalent metric (SI) pipe sizes. (*Courtesy of McGraw-Hill*)

piece of pipe weighs. You will need some information, which can be found in Figure 11.5. And, you will need the formula, which is as follows:

$$W = F \times 10.68 \times T \times (O.D. - T)$$

You're probably wondering what all the letters mean, and you should be. I'll tell you. The letter W is the weight of the pipe in pounds per foot. A relative

Nominal rod diameter, in	Root area of thread, in^2	Maximum safe load at rod temperature of 650°F, lb
¼	0.027	240
5⁄16	0.046	410
⅜	0.068	610
½	0.126	1,130
⅝	0.202	1,810
¾	0.302	2,710
⅞	0.419	3,770
1	0.552	4,960
1⅛	0.693	6,230
1¼	0.889	8,000
1⅜	1.053	9,470
1½	1.293	11,630
1⅝	1.515	13,630
1¾	1.744	15,690
1⅞	2.048	18,430
2	2.292	20,690
2¼	3.021	27,200
2½	3.716	33,500
2¾	4.619	41,600
3	5.621	50,600
3¼	6.720	60,500
3½	7.918	71,260

FIGURE 11.3 ■ Load ratings of threaded rods. (*Courtesy of McGraw-Hill*)

Pipe size, in	Rod size, in
2 and smaller	⅜
2½ to 3½	½
4 and 5	⅝
6	¾
8 to 12	⅞
14 and 16	1
18	1⅛
20	1¼
24	1½

FIGURE 11.4 ■ Recommended rod sizes for individual pipes. (*Courtesy of McGraw-Hill*)

weight factor, which can be found in Figure 11.5, is represented by the letter F. Wall thickness of a pipe is known as the letter T. You have probably guessed that O.D. represents the outside diameter of the pipe, in inches. I said that you could figure out the weight of pipe and its contents. To determine the weight of water in pipe, refer to Figure 11.6.

Pipe	Weight factor*
Aluminum	0.35
Brass	1.12
Cast iron	0.91
Copper	1.14
Stainless steel	1.0
Carbon steel	1.0
Wrought iron	0.98

*Average plastic pipe weights one-fifth as much as carbon steel pipe.

FIGURE 11.5 ■ Relative weight factors for metal pipe. (*Courtesy of McGraw-Hill*)

☞ been there **done that**

As a young plumber, I guessed at a lot of math requirements. This was not always smart. Don't gamble when it comes to pipe support. Refer to the tables here to make sure that your choice of hangers is safe and secure.

IPS, in	Weight per foot, lb	Length in feet containing 1 ft³ of water	Gallons in 1 linear ft
¼	0.42		0.005
⅜	0.57	754	0.0099
½	0.85	473	0.016
¾	1.13	270	0.027
1	1.67	166	0.05
1¼	2.27	96	0.07
1½	2.71	70	0.1
2	3.65	42	0.17
2½	5.8	30	0.24
3	7.5	20	0.38
4	10.8	11	0.66
5	14.6	7	1.03
6	19.0	5	1.5
8	25.5	3	2.6
10	40.5	1.8	4.1
12	53.5	1.2	5.9

FIGURE 11.6 ■ Weight of steel pipe and contained water. (*Courtesy of McGraw-Hill*)

THERMAL EXPANSION

Thermal expansion can occur in pipes when there are temperature fluctuations. Damage can result from this expansion if the pipe is not installed properly. In order to avoid damage, refer to Figures 11.7, 11.8, and 11.9 to learn about the tolerances needed for various types of pipe (Fig. 11.10).

Pipe material	Coefficient	
	in/in/°F	(°C)
Metallic pipe		
Carbon steel	0.000005	(14.0)
Stainless steel	0.000115	(69)
Cast iron	0.0000056	(1.0)
Copper	0.000010	(1.8)
Aluminum	0.0000980	(1.7)
Brass (yellow)	0.000001	(1.8)
Brass (red)	0.000009	(1.4)
Plastic pipe		
ABS	0.00005	(8)
PVC	0.000060	(33)
PB	0.000150	(72)
PE	0.000080	(14.4)
CPVC	0.000035	(6.3)
Styrene	0.000060	(33)
PVDF	0.000085	(14.5)
PP	0.000065	(77)
Saran	0.000038	(6.5)
CAB	0.000080	(14.4)
FRP (average)	0.000011	(1.9)
PVDF	0.000096	(15.1)
CAB	0.000085	(14.5)
HDPE	0.00011	(68)
Glass		
Borosilicate	0.0000018	(0.33)

FIGURE 11.7 ■ Thermal expansion of piping materials. (*Courtesy of McGraw-Hill*)

Length (ft)	Temperature Change (°F)						
	40	50	60	70	80	90	100
20	0.278	0.348	0.418	0.487	0.557	0.626	0.696
40	0.557	0.696	0.835	0.974	1.114	1.235	1.392
60	0.835	1.044	1.253	1.462	1.670	1.879	2.088
80	1.134	1.392	1.670	1.879	2.227	2.506	2.784
100	1.192	1.740	2.088	2.436	2.784	3.132	3.480

FIGURE 11.8 ■ Thermal expansion of PVC-DWV. (*Courtesy of McGraw-Hill*)

Length (ft)	Temperature Change (°F)						
	40	50	60	70	80	90	100
20	0.536	0.670	0.804	0.938	1.072	1.206	1.340
40	1.070	1.340	1.610	1.880	2.050	2.420	2.690
60	1.609	2.010	2.410	2.820	3.220	3.620	4.020
80	2.143	2.680	3.220	3.760	4.290	4.830	5.360
100	2.680	3.350	4.020	4.700	5.360	6.030	6.700

FIGURE 11.9 ■ Thermal expansion of all pipes (except PVC-DWV). (*Courtesy of McGraw-Hill*)

A hole in a pipe that is not more than .63 centimeters in diameter can result in a loss of 14,952 gallons of water a day! Even a pinhole leak can amount to a loss of over 18,000 gallons of water in a three-month period.

FIGURE 11.10 ■ Tech tips.

PIPE THREADS

Pipe threads come in different styles. Some are compatible, and others are not. You could encounter straight pipe threads, tapered pipe threads, or fire-hose coupling straight threads. To understand the types of pipe and hose threads, let me give you some illustrations to consider. The tables in Figures 11.11, 11.12, and 11.13 show you how many threads per inch to expect with different thread types. Fire hose threads are not compatible with any other type of threads. The same is true for garden hose threads. But, some threads are compatible with other types. If you have a female NPT thread pattern, it is compatible with male threads of an NPT type. The proper sealant to mate these threads is a thread seal. American Standard Straight Pipe (NPSM) threads on female threads can be mated to either NPSM male threads or NPT male threads. To seal such a connection, a washer seal should be used.

Pipe size (in inches)	Maximum outside diameter	Threads per inch
¼	1.375	8
1	1.375	8
1¼	1.6718	9
1½	1.990	9
2	2.5156	8
3	3.6239	6
4	5.0109	4
5	6.260	4
6	7.025	4

FIGURE 11.11 ■ Threads per inch for national standards.

Pipe size (in inches)	Maximum outside diameter	Threads per inch
¼	1.0353	14
1	1.295	11.5
1¼	1.6399	11.5
1½	1.8788	11.5
2	2.5156	8
3	3.470	8
4	4.470	8

FIGURE 11.12 ▪ **Threads per inch for American Standard Straight Pipe.**

Hose size (in inches)	Maximum outside diameter	Threads per inch
¼	1.0625	11.5

FIGURE 11.13 ▪ **Threads per inch for garden hose.**

Female threads that are NPSH can be coupled with male threads of NPSH, NPSM, or NPT types. In any of these cases, a washer seal should be used. Threads of a garden hose type are mated with a washer seal. But, what happens when you are trying to find compatible matches for a male thread pattern? If you have an NPT male thread, it can be mated to NPT, NPSM, or NPSH threads. When NPT is mated to NPT, a thread sealant should be used. Washer seals are used to mate NPSM or NPSH female threads to male NPT threads. A male NPSM thread can mate with female thread types of NPSM or NPSH. A washer seal should be used for these connections. Garden hose threads, whether male or female, can only be coupled to garden hose threads, and this is done with a washer seal.

HOW MANY TURNS?

How many turns does it take to operate a double-disk valve? It depends on the size of the valve. Refer to Figure 11.14 for the answers to how many turns

Valve size (in inches)	Number of turns required to operate valve
3	7.5
4	14.5
6	20.5
8	27
10	33.5

FIGURE 11.14 ▪ **The number of turns required to operate a double-disk valve.**

Valve size (in inches)	Number of turns required to operate valve
3	11
4	14
6	20
8	27
10	33

FIGURE 11.15 ■ **Number of turns required to operate a metal-seated sewerage valve.**

Pipe diameter (in inches)	Approximate capacity (in U.S. gallons) per foot of pipe
¾	.0230
1	.0408
1¼	.0638
1½	.0918
2	.1632
3	.3673
4	.6528
6	1.469
8	2.611
10	4.018

FIGURE 11.16 ■ **Pipe capacities.**

it takes to operate a valve. If you want to know how many turns it takes to operate a metal-seated sewerage valve, look at Figure 11.15.

PIPE CAPACITIES

Have you ever wondered what the capacity of a pipe was? You could do some heavy math to figure out the answer to your question, or you can look at Figure 11.16 for quick solutions to your questions.

WHAT IS THE DISCHARGE OF A GIVEN PIPE SIZE UNDER PRESSURE?

What is the discharge of a given pipe size under pressure? The pressure and flow are both factors to consider. If you assume that you are dealing with a straight pipe that has no bends or valves, I can give you a reference chart to use for answers to your question. Further assume that there will be open flow, with no backpressure, through a pipe with a smoothness rating of $C = 100$. Refer to Figures 11.17, 11.18, and 11.19 for quick-reference charts.

Pipe size (in inches)	PSI	Length of pipe is 50 feet
¾	20	16
¾	40	24
¾	60	29
¾	80	34
1	20	31
1	40	44
1	60	55
1	80	65
1¼	20	84
1¼	40	121
1¼	60	151
1¼	80	177
1½	20	94
1½	40	137
1½	60	170
1½	80	200

FIGURE 11.17 ■ Discharge of pipes in gallons per minute.

Pipe size (in inches)	PSI	Length of pipe is 100 feet
¾	20	11
¾	40	16
¾	60	20
¾	80	24
1	20	21
1	40	31
1	60	38
1	80	44
1¼	20	58
1¼	40	84
1¼	60	104
1¼	80	121
1½	20	65
1½	40	94
1½	60	117
1½	80	137

FIGURE 11.18 ■ Discharge of pipes in gallons per minute.

Pipe size (in inches)	PSI	Length of pipe is 200 feet
¾	20	8
¾	40	11
¾	60	14
¾	80	16
1	20	14
1	40	21
1	60	26
1	80	31
1¼	20	39
1¼	40	58
1¼	60	72
1¼	80	84
1½	20	45
1½	40	65
1½	60	81
1½	80	94

FIGURE 11.19 ■ **Discharge of pipes in gallons per minute.**

SOME FACTS ABOUT COPPER PIPE AND TUBING

Would you like some facts about copper pipe and tubing? Well, you're in the right place. Let's go over some data that could serve you well in your plumbing endeavors. Figure 11.20 will show you some size data for copper tubing. Are you interested in size details for copper that is used for drain, waste, and vent (DWV) applications? Refer to Figure 11.21 for this information.

Copper is rated in terms of types. For example, Type K copper has a thick wall and is considered a stronger material than Type L or Type M copper. This type of tubing isn't used often in residential work, but it is sometimes used for water services when the copper is supplied in its soft form. Soft copper comes in a roll and allows underground piping, such as that for a water service, to be installed without joints. Type L copper is frequently used for water distribution pipes in homes and can be used in its soft form for water services. A softer type of copper is known as Type M copper. This copper tubing is used mostly for hot-water-baseboard heating systems. It is available only in rigid lengths and is not available in a rolled coil. Many plumbing codes prohibit its use for water distribution systems. Figure 11.22 will show you how different types of copper are available for purchase.

▶ *sensible* **shortcut**

Use Type L copper for water distribution piping. Even if your local code allows Type M copper, Type L copper is better, will last longer, and is less susceptible to pinhole leaks than Type M copper. The cost difference is noticeable, but the peace of mind that you have in knowing that your customers will be happy longer is invaluable.

Nominal pipe size (inches)	Outside diameter (inches)	Inside diameter (inches)
Type K		
¼	0.375	0.305
⅜	0.500	0.402
½	0.625	0.527
⅝	0.750	0.652
¾	0.875	0.745
1	1.125	0.995
1¼	1.375	1.245
1½	1.625	1.481
2	2.125	1.959
2½	2.625	2.435
3	3.125	2.907
3½	3.625	3.385
4	4.125	3.857
5	5.125	4.805
6	6.125	5.741
8	8.125	7.583
10	10.125	9.449
12	12.125	11.315
Type L		
¼	0.375	0.315
⅜	0.500	0.430
½	0.625	0.545
⅝	0.750	0.666
¾	0.875	0.785
1	1.125	1.025
1¼	1.375	1.265
1½	1.625	1.505
2	2.125	1.985
2½	2.625	2.465
3	3.125	2.945
3½	3.625	3.425
4	4.125	3.905
5	5.125	4.875
6	6.125	5.845
8	8.125	7.725
10	10.125	9.625
12	12.125	11.565

FIGURE 11.20 ■ Copper tube - water distribution.
(*Courtesy of McGraw-Hill*)

Inside diameter (inches)	Nominal size (inches)	Outside diameter (inches)
Type DWV		
N/A	¼	0.375
N/A	⅜	0.500
N/A	½	0.625
N/A	⅝	0.750
N/A	¾	0.875
N/A	1	1.125
1.295	1¼	1.375
1.511	1½	1.625
2.041	2	2.125
3.035	2½	2.625
N/A	3	3.125
N/A	3½	3.625
4.009	4	4.125
4.981	5	5.125
5.959	6	6.125
N/A	8	8.125
N/A	10	10.125
N/A	12	12.125

FIGURE 11.21 ■ Copper tube - DWV. (*Courtesy of McGraw-Hill*)

CAST IRON

Cast-iron pipe comes in three basic types. One is known as service-weight pipe and another is called extra-heavy cast iron. These types of pipe may be purchased with either one or two hubs. A third type of cast-iron pipe is called no-hub pipe. This type has no hub on either end; it is coupled with mechanical joints (Fig. 11.24 & 11.25). Cast iron is still in use and provides years of dependable service.

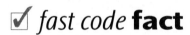
✓ *fast code* **fact**

Don't use 50-50 solder for potable water piping. Most codes require lead-free solder or, at the most, a 95-5 solder for potable water piping. Solder with a 50-50 rating is normally used only for heating pipes at this stage of life.

PLASTIC PIPE FOR DRAINS & VENTS

Plastic pipe for drains and vents is very common in modern plumbing. Polyvinyl Chloride Plastic Pipe (PVC) is probably used more often than any other type of drainage or vent pipe (Fig. 11.26). This type of pipe is strong and resistive to a variety of acids and bases. PVC pipe can be used with water, gas, and drainage systems, but it is not rated for use with hot water. I've found this type of pipe to be sensitive to dirt and water when joints are being

Drawn (hard copper) (feet)		Annealed (soft copper) (feet)	
Type K Tube			
Straight Lengths:		Straight Lengths:	
Up to 8-in. diameter	20	Up to 8-in. diameter	20
10-in. diameter	18	10-in. diameter	18
112-in. diameter	12	12-in. diameter	12
		Coils:	
		Up to 1-in. diameter	60
		1¼-in. diameter	60
			100
		2-in. diameter	40
			45
Type L Tube			
Straight Lengths:		Straight Lengths:	
Up to 10-in. diameter	20	Up to 10-in. diameter	20
12-in. diameter	18	12-in. diameter	18
		Coils:	
		Up to 1-in. diameter	60
			100
		1¼- and 1½-in. diameter	60
			100
		2-in. diameter	40
			45
DWV Tube			
Straight Lengths:		Not available	
All diameters	20		
Type M Tube			
Straight Lengths:		Not available	
All diameters	20		

FIGURE 11.22 ■ Available lengths of copper plumbing tube. (*Courtesy of McGraw-Hill*)

Hard copper is also known as drawn copper, while soft copper tubing is known as annealed copper.

FIGURE 11.23 ■ Tech tips.

made. The areas being joined should be dry, clean, and primed prior to solvent welding. Also, PVC becomes brittle in cold weather and should not be dropped on hard surfaces.

Acrylonitrile Butadiene Styrene (ABS) pipe is the drainage pipe of preference for me. However, I do use more PVC than ABS at this point in my

Size (inches)	Service weight per linear foot (pounds)	Extra heavy size (inches)	Per linear foot (pounds)
2	4	2	5
3	6	3	9
4	9	4	12
5	12	5	15
6	15	6	19
7	20	8	30
8	25	10	43
		12	54
		15	75

FIGURE 11.24 ■ Weight of cast-iron soil and pipe. (*Courtesy of McGraw-Hill*)

career. When plastic drainage and vent piping became popular, I cut my teeth on ABS pipe. But, PVC pipe is less expensive in most regions and enjoys a less-destructive rating in the case of fires, so most of the industry, that I know of, has moved to PVC. I like ABS because of its durability and its ease of working with. This pipe is so strong that I've seen loaded dump trucks run over a section of ABS on a project and never crush the pipe!

☑ *fast code* **fact**

Most local codes require the application of a purple primer when PVC pipe is installed. The purple color allows inspectors to see that a cleaner/primer has been used. This type of primer is messy and many homeowners complain about it, but it is often required by the code enforcement office.

PIPING COLOR CODES

Piping color codes are used when utility companies stake out piping locations. For example, a yellow flag generally indicates one of the following types of pipe:

- Oil
- Steam
- Gas
- Petroleum

When you encounter a blue flag that indicates the location of piping below ground, the type of piping that you are probably dealing with is one of the following:

- Potable Water
- Irrigation Water
- Slurry Pipes

	Diameter (inches)	Service weight (lb)	Extra heavy weight (lb)
Double hub, 5-ft lengths	2	21	26
	3	31	47
	4	42	63
	5	54	78
	6	68	100
	8	105	157
	10	150	225
Double hub, 30-ft length	2	11	14
	3	17	26
	4	23	33
Single hub, 5-ft lengths	2	20	25
	3	30	45
	4	40	60
	5	52	75
	6	65	95
	8	100	150
	10	145	215
Single hub, 10-ft lengths	2	38	43
	3	56	83
	4	75	108
	5	98	133
	6	124	160
	8	185	265
	10	270	400
No-hub pipe, 10-ft lengths	1½	27	
	2	38	
	3	54	
	4	74	
	5	95	
	6	118	
	8	180	

FIGURE 11.25 ■ Weight of cast-iron pipe. (*Courtesy of McGraw-Hill*)

Green flags tend to mark the locations of sewers and drain lines. You can never count on the colors to be right and you should always check with the flagging company to know what types of pipes you may be dealing with, but the above examples are common choices when color-coded flags are used. Now, let's go to Chapter 12 and see how you can troubleshoot jobs by using tables and common sense for fast solutions to serious problems.

Nominal pipe size (inches)	Outside diameter (inches)	Inside diameter (inches)	Wall thickness (inches)
½	0.840	0.750	0.045
¾	1.050	0.940	0.055
1	1.315	1.195	0.060
1¼	1.660	1.520	0.070
1½	1.900	1.740	0.080
2	2.375	2.175	0.100
2½	2.875	2.635	0.120
3	3.500	3.220	0.140
4	4.500	4.110	0.195

FIGURE 11.26 ■ Polyvinyl Chloride Plastic Pipe (PVC). (*Courtesy of McGraw-Hill*)

chapter 12

TROUBLESHOOTING

Figures 12.1 through 12.27 provide useful tables to help you in troubleshooting problems.

1. What well conditions might possibly limit the capacity of the pump?	The rate of flow from the source of supply, the diameter of a cased deep well, and the pumping level of the water in a cased deep well.
2. How does the diameter of a cased deep well and pumping level of the water affect the capacity?	They limit the size pumping equipment which can be used.
3. If there are no limiting factors, how is capacity determined?	By the maximum number of outlets or faucets likely to be in use at the same time.
4. What is suction?	A partial vacuum, created in the suction chamber of the pump, obtained by removing pressure due to atmosphere, thereby allowing greater pressure outside to force something (air, gas, water) into the container.
5. What is atmospheric pressure?	The atmosphere surrounding the earth presses against the earth and all objects on it, producing what we call atmospheric pressure.
6. How much is the pressure due to atmosphere?	This pressure varies with elevation or altitude. It is greatest at sea level (14.7 pounds per square inch) and gradually decreases as elevation above sea level is increased. The rate is approximately 1 foot per 100 feet of elevation.
7. What is maximum theoretical suction lift?	Since suction lift is actually that height to which atmospheric pressure will force water into a vacuum, theoretically we can use the maximum amount of this pressure 14.7 pounds per square inch at sea level which will raise water 33.9 feet. From this, we obtain the conversion factor of 1 pound per square inch of pressure equals 2.31-feet head.
8. How does friction loss affect suction conditions?	The resistance of the suction pipe walls to the flow of water uses up part of the work which can be done by atmospheric pressure. Therefore, the amount of loss due to friction in the suction pipe must be added to the vertical elevation which must be overcome, and the total of the two must not exceed 25 feet at sea level. This 25 feet must be reduced 1 foot for every 1000-feet elevation above sea level, which corrects for a lessened atmospheric pressure with increased elevation.

FIGURE 12.1 ▪ Questions and answers about pumps. (*Courtesy of McGraw-Hill*)

9. When and why do we use a deep-well jet pump?

The resistance of the suction pipe walls to below the pump because this is the maximum practical suction lift which can be obtained with a shallow-well pump at sea level.

10. What do we mean by water systems?

A pump with all necessary accessories, fittings, etc., necessary for its completely automatic operation.

11. What is the purpose of a foot value?

It is used on the end of a suction pipe to prevent the water in the system from running back into the source of supply when the pump isn't operating.

12. Name the two basic parts of a jet assembly.

Nozzle and diffuser.

13. What is the function of the nozzle?

The nozzle converts the pressure of the driving water into velocity. The velocity thus created causes a vacuum in the jet assembly or suction chamber.

14. What is the purpose of the diffuser?

The diffuser converts the velocity from the nozzle back into pressure.

15. What do we mean by "driving water"?

That water which is supplied under pressure to drive the jet.

16. What is the source of the driving water?

The driving water is continuously recirculated in a closed system.

17. What is the purpose of the centrifugal pump?

The centrifugal pump provides the energy to circulate the driving water. It also boosts the pressure of the discharged capacity.

18. Where is the jet assembly usually located in a shallow-well jet system?

Bolted to the casing of the centrifugal pump.

19. What is the principal factor which determines if a shallow-well jet system can be used?

Total suction lift.

20. When is a deep-well jet system used?

When the total suction sift exceeds that which can be overcome by atmospheric pressure.

21. Can a foot valve be omitted from a deep-well jet system? Why or why not?

No, because there are no valves in the jet assembly, and the foot valve is necessary to hold water in the system when it is primed. Also, when the centrifugal pump isn't running, the foot valve prevents the water from running back into the well.

FIGURE 12.1 ■ (Continued) Questions and answers about pumps. (Courtesy of McGraw-Hill)

22. What is the function of a check valve in the top of a submersible pump?	To hold the pressure in the line when the pump isn't running.
23. A submersible pump is made up of two basic parts. What are they?	Pump end and motor.
24. Why did the name submersible pump come into being?	Because the whole unit, pump and motor, is designed to be operated under water.
25. Can a submersible pump be installed in a 2-inch well?	No, they require a 4-inch well or larger for most domestic use. Larger pumps with larger capacities require 6-inch wells or larger.
26. A stage in a submersible pump is made up of three parts. What are they?	Impeller, diffuser, and bowl.
27. Does a submersible pump have only one pipe connection?	Yes, the discharge pipe.
28. What are two reasons we should always consider using a submersible first?	It will pump more water at higher pressure with less horsepower. It also has easier installation.
29. The amount of pressure a pump is capable of making is controlled by what?	The diameter of the impeller.
30. What do the width of an impeller and guide vane control?	The amount of water or capacity the pump is capable of pumping.

FIGURE 12.1 ■ (*Continued*) Questions and answers about pumps. (*Courtesy of McGraw-Hill*)

Motor does not start		
Cause of trouble	Checking procedure	Corrective action
No power or incorrect voltage.	Using voltmeter, check the line terminals. Voltage must be ± 10% of rated voltage.	Contact power company if voltage is incorrect.
Fuses blown or circuit fuse breakers tripped.	Check fuses for recommended size and check for loose, dirty, or corroded connections in fuse receptacle. Check for tripped circuit breaker.	Replace with proper or reset circuit breaker.
Defective pressure switch.	Check voltage at contact points. Improper contact of switch points can cause voltage less than line voltage.	Replace pressure switch or clean points.
Control box malfunction.	For detailed procedure, see***	Repair or replace.
Defective wiring.	Check for loose or corroded connections. Check motor lead terminals with voltmeter for power.	Correct faulty wiring or connections.
Bound pump.	Locked rotor conditions can result from misalignment between pump and motor or a sand bound pump. Amp readings 3 to 6 times higher than normal will be indicated.	If pump will not start with several trials, it must be pulled and the cause corrected. New installations should always be run without turning off until water clears.
Defective cable or motor.		Repair or replace.
Motor starts too often		
Pressure switch.	Check setting on pressure switch and examine for defects.	Reset limit or replace switch.
Check valve, stuck open.	Damaged or defective check valve will not hold pressure.	Replace if defective.
Waterlogged tank (air supply).	Check air-charging system for proper operation.	Clean or replace.
Leak in system.	Check system for leaks.	Replace damaged pipes or repair leaks.

SOURCE: A. Y. McDonald Manufacturing Co.

FIGURE 12.2 ▪ Troubleshooting motors. (*Courtesy of McGraw-Hill*)

Motor runs continuously		
Causes of trouble	Checking procedure	Corrective action
Pressure switch.	Switch contacts may be "welded" in closed position. Pressure switch may be set too high.	Clean contacts, replace switch, or readjust setting.
Low-level well.	Pump may exceed well capacity. Shut off pump, wait for well to recover. Check static and draw-down level from well head.	Throttle pump output or reset pump to lower level. Do not lower if sand may clog pump.
Leak in system.	Check system for leaks.	Replace damaged pipes or repair leaks.
Worn pump.	Symptoms of worn pump are similar to that of drop pipe leak or low water level in well. Reduce pressure switch setting. If pump shuts off, worn parts may be at fault. Sand is usually present in tank.	Pull pump and replace worn impellers, casing, or other close fitting parts.
Loose or broken motor shaft.	No or little water will be delivered if coupling between motor and pump shaft is loose or if a jammed pump has caused the motor shaft to shear off.	Check for damaged shafts if coupling is loose, and replace worn or defective units.
Pump screen blocked.	Restricted flow may indicate a clogged intake screen on pump. Pump may be installed in mud or sand.	Clean screen and reset at less depth. It may be necessary to clean well.
Check valve stuck closed.	No water will be delivered if check valve is in closed position.	Replace if defective.
Control box malfunction.		Repair or replace.
Motor runs but overload protector tips		
Incorrect voltage.	Using voltmeter, check the line terminals. Voltage must be within ± 10% of rated voltage.	Contact power company if voltage is incorrect.
Overheated protectors.	Direct sunlight or other heat source can make control box hot, causing protectors to trip. The box must not be hot to touch.	Shade box, provide ventilation, or move box away from heat source.
Defective control box.		Repair or replace.
Defective motor or cable.		Repair or replace.
Worn pump or motor.		Replace pump and/or motor.

SOURCE: A. Y. McDonald Manufacturing Co.

FIGURE 12.3 ■ Troubleshooting motors. (*Courtesy of McGraw-Hill*)

Cable size	Resistance
14	0.5150
12	0.3238
10	0.2036
8	0.1281
6	0.08056
4	0.0506
2	0.0318

FIGURE 12.4 ■ Resistance
of electrical wire.
(*Courtesy of McGraw-Hill*)

Cause of trouble	Checking procedure	Correction action
Motor does not start		
A. No power or incorrect voltage.	Using voltmeter check the line terminals Voltage must be ±10% of rated voltage.	Contact power company if voltage is incorrect.
B. Fuses blown or circuit breakers tripped.	Check fuses for recommended size and check for loose, dirty or corroded connections in fuse receptacle. Check for tripped circuit breaker.	Replace with proper fuse or reset circuit breaker.
C. Defective pressure switch.	Check voltage at contact points. Improper contact of switch points can cause voltage less than line voltage.	Replace pressure switch or clean points.
D. Control box malfunction.		Repair or replace.
E. Defective wiring.	Check for loose or corroded connections. Check motor lead terminals with voltmeter for power.	Correct faulty wiring or connections.
F. Bound pump.	Locked rotor conditions can result from misalignment between pump and motor or a sand bound pump. Amp readings 3 to 6 times higher than normal will be indicated.	If pump will not start with several trials it must be pulled and the cause corrected. New installations should always be run without turning off until water clears.
G. Defective cable or motor.		Repair or replace.
Motor starts too often		
A. Pressure switch.	Check setting on pressure switch and examine for defects.	Reset limit or replace switch.
B. Check valve, stuck open.	Damaged or defective check valve will not hold pressure.	Replace if defective.
C. Waterlogged tank, (air supply)	Check air charging system for proper operation.	Clean or replace.
D. Leak in system.	Check system for leaks.	Replace damaged pipes or repair leaks.

FIGURE 12.5 ■ Troubleshooting motors. (*Courtesy of McGraw-Hill*)

Motor runs continuously		
Cause of trouble	Checking procedure	Correction action
A. Pressure switch.	Switch contacts may be "welded" in closed position. Pressure switch may be set too high.	Clean contacts replace switch, or readjust setting.
B. Low level well.	Pump may exceed well capacity. Shut off pump, wait for well to recover. Check static and drawdown level from well head.	Throttle pump output or reset pump to lower level. Do not lower if sand may clog pump.
C. Leak in system.	Check system for leaks.	Replace damaged pipes or repair leaks.
D. Worn pump, motor shaft.	Symptoms of worn pump are similar to those of drop pipe leak or low water level in well. Reduce pressure switch setting. If pump shuts off, worn parts may be at fault. Sand is usually present in tank.	Pull pump and replace worn impellers, casing or other close fitting parts.
E. Loose or broken.	No or little water will be delivered if coupling between motor and pump shaft is loose or if a jammed pump has caused the motor shaft to shear off.	Check for damaged shafts if coupling is loose and replace worn or defective units.
F. Pump screen blocked.	Restricted flow may indicate a clogged intake screen on pump. Pump may be installed in mud or sand.	Clean screen and reset at less depth. It may be necessary to clean well.
G. Check valve stuck closed.	No water will be delivered if check valve is in closed position.	Replace if defective.
H. Control box malfunction.		Repair or replace.
Motor runs but overload protector trips		
A. Incorrect voltage.	Using voltmeter, check the line terminals. Voltage must be within ± 10% of rated voltage.	Contact power company if voltage is incorrect.
B. Overheated protectors.	Direct sunlight or other heat source can make control box hot causing protectors to trip. The box must not be hot to touch.	Shade box, provide ventilation or move box away from heat source.
C. Defective control box.		Repair or replace.
D. Defective motor or cable.		Repair or replace.
E. Worn pump or motor.		Replace pump and/or motor.

FIGURE 12.6 ▪ Troubleshooting motors. (*Courtesy of McGraw-Hill*)

Preliminary tests—all sizes—single and three phase	
What is to be done	What it means
Measure resistance from any cable to ground (insulation resistance)	1. If the ohm value is normal, the motor windings are not grounded and the cable insulation is not damaged.
	2. If the ohm value is below normal, either the windings are grounded or the cable insulation is damaged. Check the cable at the well seal as the insulation is sometimes damaged by being pinched.
Measure winding resistance (resistance between leads)	1. If all ohm values are normal, the motor windings are neither shorted nor open, and the cable colors are correct.
	2. If any one ohm value is less than normal, the motor is shorted.
	3. If any one ohm value is greater than normal, the winding or the cable is open, or there is a poor cable joint or connection.
	4. If some ohm values are greater than normal and some less on single phase motors, the leads are mixed.

FIGURE 12.7 ■ Troubleshooting motors. (*Courtesy of McGraw-Hill*)

Normal ohm and megohm values between all leads and ground		
Insulation resistance varies very little with rating. Motors of all hp, voltage, and phase rating have similar values of insulation resistance.		
Condition of motor and leads	Ohm value	Megohm value
A new motor (without drop cable).	20,000,000 (or more)	20.0 (or more)
A used motor which can be reinstalled in the well.	10,000,000 (or more)	10.0 (or more)
Motor in well. Ohm readings are for drop cable plus motor.		
A new motor in the well.	2,000,000 (or more)	2.0 (or more)
A motor in the well in reasonably good condition.	500,000–2,000,000	0.5–2.0
A motor which may have been damaged by lightning or with damaged leads. Do not pull the pump for this reason.	20,000–500,000	0.02–0.5
A motor which definitely has been damaged or with damaged cable. The pump should be pulled and repairs made to the cable or motor replaced. The motor will not fail for this reason alone, but it will probably not operate for long.	10,000–20,000	0.01–0.02
A motor which has failed or with completely destroyed cable insulation. The pump must be pulled and the cable repaired or the motor replaced.	less than 10,000	0–0.01

FIGURE 12.8 ■ Resistance readings. (*Courtesy of McGraw-Hill*)

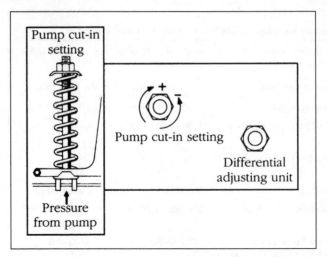

FIGURE 12.9 ▪ **Fine-tuning instructions for pressure switches. (*Courtesy of McGraw-Hill*)**

Meter connections for motor testing

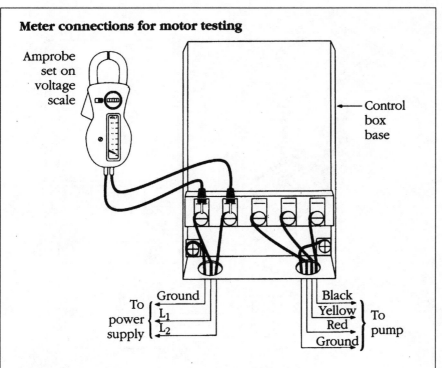

Amprobe set on voltage scale

Control box base

To power supply { Ground, L₁, L₂ }

To pump { Black, Yellow, Red, Ground }

To check voltage

1. Turn power OFF

2. Remove QD cover to break all motor connections.

Caution: L1 and L2 are still connected to the power supply.

3. Turn power ON.

4. Use voltmeter as shown.

Caution: Both voltage and current tests require live circuits with power ON.

FIGURE 12.10 ■ Meter connections for motor testing. (*Courtesy of McGraw-Hill*)

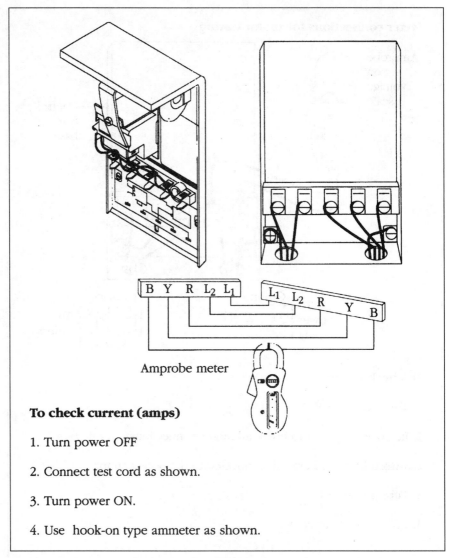

Amprobe meter

To check current (amps)

1. Turn power OFF

2. Connect test cord as shown.

3. Turn power ON.

4. Use hook-on type ammeter as shown.

FIGURE 12.11 ■ Checking amperage. (*Courtesy of McGraw-Hill*)

Single-phase control boxes

**Checking and repairing procedures
(Power on)**

Caution: Power must be on for these tests. Do not touch any live parts.

A. General procedures:
   1. Establish line power.
   2. Check no load voltage (pump not running).
   3. Check load voltage (pump running).
   4. Check current (amps) in all motor leads.

B. Use of volt/amp meter:
   1. Meter such as Amprobe Model RS300 or equivalent may be used.
   2. Select scale for voltage or amps depending on tests.
   3. When using amp scales, select highest scale to allow for inrush
      current, then select for midrange reading.

C. Voltage measurements:
   Step 1, no load.
   1. Measure voltage at L1 and L2 of pressure switch or line contractor.
   2. Voltage Reading: Should be ±10% of motor rating.
   Step 2, load.
   1. Measure voltage at load side of pressure switch or line contractor with
      pump running.
   2. Voltage Reading: Should remain the same except for slight dip on
      starting.

D. Current (amp) measurements:
   1. Measure current on all motor leads. Use 5 conductor test cord for Q.D.
      control boxes.
   2. Amp Reading: Current in Red lead should momentarily be high, then
      drop within one second. This verifies relay or solid state relay
      operation.

E. Voltage symptoms:
   1. Excessive voltage drop on starting.
   2. Causes: Loose connections, bad contacts or ground faults, or
      inadequate power supply.

F. Current symptoms:
   1. Relay or switch failures will cause Red lead current to remain high
      and overload tripping.
   2. Open.run capacitor(s) will cause amps to be higher than normal in the
      Black and Yellow motor leads and lower than normal or zero amps in
      the Red motor lead.
   3. Relay chatter is caused by low voltage or ground faults.
   4. A bound pump will cause locked rotor amps and overloading tripping.
   5. Low amps may be caused by pump running at shutoff, worn pump or
      stripped splines.
   6. Failed start capacitor or open switch/relay are indicated if the red lead
      current is not momentarily high at starting.

FIGURE 12.12 ■ Troubleshooting motors. (*Courtesy of McGraw-Hill*)

Single-phase control boxes

Checking and repairing procedures
(Power off)

Caution: Turn power off at the power supply panel and discharge capacitors before using ohmmeter.

A. General procedures:
 1. Disconnect line power.
 2. Inspect for damaged or burned parts, loose connections, etc.
 3. Check against diagram in control box for misconnections.
 4. Check motor insulation and winding resistance.

B. Use of ohmmeter:
 1. Ohmmeter such as Simpson Model 372 or 260. Triplet Model 630 or 666 may be used.
 2. Whenever scales are changed, clip ohmmeter lead together and "zero balance" meter.

C. Ground (insulation resistance) test:
 1. Ohmmeter Setting: Highest scale R × 10K, or R × 100K
 2. Terminal Connections: One ohmmeter lead to "Ground" terminal or Q.D. control box lid and touch other lead to the other terminals on the terminal board.
 3. Ohmmeter Reading: Pointer should remain at infinity (∞).

Additional tests

Solid state capacitor run
(CRC) control box

A. Run capacitor
 1. Meter setting: R × 1,000
 2. Connections: Red and Black leads
 3. Correct meter reading: Pointer should swing toward zero, then drift back to infinity.

B. Inductance coil
 1. Meter setting: R × 1
 2. Connections: Orange leads
 3. Correct meter reading: Less than 1 ohm.

C. Solid state switch
 Step 1 triac test
 1. Meter setting: R × 1,000
 2. Connections: R(Start) terminal and Orange lead on start switch.
 3. Correct meter reading: Should be near infinity after swing.
 Step 2 coil test
 1. Meter setting: R × 1
 2. Connections: Y(Common) and L2.
 3. Correct meter reading: Zero ohms

FIGURE 12.13 ■ Troubleshooting motors. (*Courtesy of McGraw-Hill*)

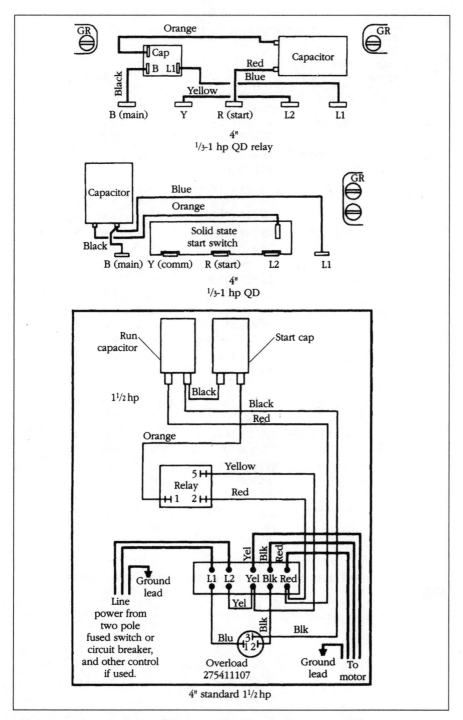

FIGURE 12.14 ■ Wiring diagrams. (*Courtesy of McGraw-Hill*)

Integral horsepower control box parts					
Motor rating hp dia.	Control box (1) model no.	Part no. (2)	Capacitors MFD	Volts	Qty.
5–6"	282 2009 202	275 468 117 S	130–154	330	2
		275 479 103 (5)	15	370	2
	282 2009 203	275 468 117 S	130–154	330	2
		155 327 101 R	30	370	1
5–6" DLX	282 2009 303	275 468 117 S	130–154	330	2
		155 327 101 R	30	370	1
7½–6"	282 2019 210	275 468 119 S	270–324	330	1
		275 468 117 S	130–154	330	1
		155 327 109 R	45	370	1
	282 2019 202	275 468 117S	130–154	330	3
		275 479 103 R (5)	15	370	3
	282 2019 203	275 468 117 S	130–154	330	3
		155 327 101 R	30	370	1
		155 328 101 R	15	370	1
7½–6" DLX	282 2019 310	275 468 119 S	270–324	330	1
		275 468 117 S	130–154	330	1
		155 327 109 R	45	370	1
	282 2019 303	275 468 117 S	130–154	330	3
		155 327 101 R	30	370	1
		155 328 101 R	15	370	1
10–6"	282 2029 210	275 468 119 S	270–324	330	2
		155 327 102 R	35	370	2
	282 2029 202	275 468 117 S	130–154	330	4
		275 479 103 R (5)	15	370	5
	282 2029 203	275 468 117 S	130–154	330	4
		155 327 101 R	30	370	2
		155 328 101 R	15	370	1
	282 2029 207	275 468 119 S	270–324	330	2
		155 327 101 R	30	370	2
		155 328 101 R	15	370	1

FIGURE 12.15 ■ Data chart for single-phase motors. (*Courtesy of McGraw-Hill*)

10–6″ DLX	282 2029 310	275 468 119 S	270–324	330	2	
		155 327 102 R	35	370	2	
	282 2029 303	275 468 117 S	130–154	330	4	
		155 327 101 R	30	370	2	
		155 328 101 R	15	370	1	
	282 2029 307	275 468 119 S	270–324	330	2	
		155 327 101 R	30	370	2	
		155 328 101 R	15	370	1	
15–6″ DLX	282 2039 310	275 468 119 S	270–324	330	2	
		155 327 109 R	45	370	3	
	282 2039 303	275 468 119 S	270–324	330	2	
		155 327 101 R	30	370	4	
		155 328 101 R	15	370	1	

FOOTNOTES:
(1) Lightning arrestor 150 814 902 suitable for all control boxes
(2) S = Start M = Main L = Line R = Run DXL = Deluxe control box with line contractor.
(3) Capacitor and overload assembly.
(4) 2 required
(5) These parts may be replaced as follows:

Old	New
275 479 102	155 328 102
275 479 103	155 328 101
275 479 105	155 328 103
275 481 102	155 327 102

FIGURE 12.15 ■ (*Continued*) Data chart for single-phase motors. (*Courtesy of McGraw-Hill*)

Integral horsepower control box parts					
Motor rating hp dia.	Control box (1) model no.	Part no.	Capacitors MFD	Volts	Qty.
1½–4"	282 3008 110	275 464 113 S	105–126	220	1
		155 328 102 R	10	370	1
	282 3007 202 or	275 461 107 S	105–126	220	1
	282 3007 102	275 479 102 R (5)	10	370	1
	282 3007 203 or	275 461 107 S	105–126	220	1
	282 3007 103	155 328 102 R	10	370	1
2–4"	282 3018 110	275 464 113 S	105–126	220	1
		155 328 103 R	20	370	1
	282 3018 202	275 464 113 S	105–126	220	1
		275 479 105 R (5)	20	370	1
	282 3018 203 or	275 464 113 S	105–126	220	1
	282 3018 103	155 328 103 R	20	370	1
2–4" DLX	282 3018 310	275 464 113 S	105–126	220	1
		155 328 103 R	20	370	1
	282 3019 103	275 464 113 S	105–126	220	1
		155 328 103 R	20	370	1
3–4"	282 3028 110	275 463 111 S	208–250	220	1
		155 327 102 R	35	370	1
	282 3028 202	275 463 111 S	208–250	220	1
		275 481 102 R (5)	35	370	2
	282 3028 203 or	275 463 111 S	208–250	220	1
	282 3028 103	155 327 102 R	35	370	1
3–4" DLX	282 3028 310	275 463 111 S	208–250	220	1
		155 327 102 R	35	370	1
	282 3029 103	275 463 111 S	208–250	220	1
		155 327 102 R	35	370	1
5–4" & 6"	282 1138 110	275 468 118 S	216–259	330	1
		155 327 101 R	30	370	2
5–4"	282 1139 202	275 468 118 S	216–259	330	1
		275 479 103 R (5)	15	370	4
	282 1139 203 or	275 468 118 S	216–259	330	1
	282 1139 003	155 327 101 R	30	370	2
5–4" & 6" DLX	282 1138 310 or	275 468 118 S	216–259	330	1
	282 1139 310	155 327 101 R	30	370	2
5–4" DLX	282 1139 303 or	275 468 118 S	216–259	330	1
	282 1139 103	155 327 101 R	30	370	2

FOOTNOTES:
(1) Lightning arrestor 150 814 902 suitable for all control boxes
(2) S = Start M = Main L = Line R = Run DXL = Deluxe control box with line contactor.
(3) Capacitor and overload assembly.
(4) 2 required
(5) These parts may be replaced as follows:

Old	New
275 479 102	155 328 102
275 479 103	155 328 101
275 479 105	155 328 103
275 481 102	155 327 102

FIGURE 12.16 ■ Data chart for single-phase motors. (*Courtesy of McGraw-Hill*)

			QD control box parts		
Hp	Volts	Control box model no.	(1) Solid state SW or QD (blue) relay	Start capacitor	MFD
⅓	115	2801024910	152138905(5)	275464125	159–191
		2801024915	223415905(5)	275464125	159–191
⅓	230	2801034910	152138901(5)	275464126	43–53
		2801034915	223415901(5)	275464126	43–53
½	115	2801044910	152138906(5)	275464201	250–300
		2801044915	223415906(5)	275464201	250–300
½	230	2801054910	152138902(5)	275464105	59–71
		2801054915	223415902(5)	275464105	59–71
½	230	2824055010	152138912	275470115	43–52
		2824055015	223415912(6)	275464105	59–71
¾	230	2801074910	152138903(5)	275464118	86–103
		2801074915	223415903(5)	275464118	86–103
¾	230	2824075010	152138913	275470114	108–130
		2824075015	223415913(6)	275470114	86–103
1	230	2801084910	152138904(5)	275464113	105–126
		2801084915	223415904(5)	275464113	105–126
1	230	2824085010	152138914	275470114	108–130
		2824085015	223415914(6)	275470114	108–130

FOOTNOTES:

(1) Prefixes 152 are solid state switches. Prefixes 223 are QD (Blue) Relays.

(2) Control boxes supplied with solid state relays are designed to operate on normal 230 V systems. For 208 V systems or where line voltage is between 200 V use the next larger cable size, or use boost transformer to raise the voltage to 230 V.

(3) Voltage relay kits 115 V, 305 102 901 and 230 V. 305 102 902 will replace either current voltage or QD Relays, and solid state switches.

(4) QD control boxes produced H85 or later do not contain an overload in the capacitor. On winding thermal overloads were added to three-wire motors rated ⅓-1 hp in A85. If a control box dated H85 or later is applied with a motor dated M84 or earlier, overload protection can be provided by adding an overload kit to the control box.

(5) May be replaced with QD relay kits 305 101 901 thru 906. Use same kit suffix as switch or relay suffix.

(6) Replace with CRC QD Relaying Kits, 223 415 912 with 305 105 901, 223 415 913 with 305 105 902 and 223 415 914 with 305 105 903.

FIGURE 12.17 ▪ Data chart for single-phase motors. (*Courtesy of McGraw-Hill*)

Symptoms	Probable cause
Won't start	No electrical power Wrong voltage Bad pressure switch Bad electrical connection Bad motor Motor contacts are open Motor shaft is seized
Runs, but produces no water	Needs to be primed Foot valve is above the water level in the well Strainer clogged Suction leak
Starts and stops too often	Leak in the piping Bad pressure switch Bad air control valve Waterlogged pressure tank Leak in pressure tank
Low water pressure in pressure tank	Strainer on foot valve is partially blocked Leak in piping Bad air charger Worn impeller hub Lift demand is too much for the pump
Pump does not cut off when working pressure is obtained	Pressure switch is bad Pressure switch needs adjusting Blockage in the piping

FIGURE 12.18 ■ Troubleshooting jet pumps. (*Courtesy of McGraw-Hill*)

Symptoms	Probable cause
Won't start	No electrical power
	Wrong voltage
	Bad pressure switch
	Bad electrical connection
Starts, but shuts off fast	Circuit breaker or fuse is inadequate
	Wrong voltage
	Bad control box
	Bad electrical connections
	Bad pressure switch
	Pipe blockage
	Pump is seized
	Control box is too hot
Runs, but does not produce water, or produces only a small quantity	Check valve stuck in closed position
	Check valve installed backward
	Bad electrical wiring
	Wrong voltage
	Pump is sitting above the water in the well
	Leak in the piping
	Bad pump or motor
	Broken pump shaft
	Clogged strainer
	Jammed impeller
Low water pressure in pressure tank	Pressure switch needs adjusting
	Bad pump
	Leak in piping
	Wrong voltage
Pump runs too often	Check valve stuck open
	Pressure tank is waterlogged and needs air injected
	Pressure switch needs adjusting
	Leak in piping
	Wrong-size pressure tank

FIGURE 12.19 ■ Troubleshooting submersible potable-water pumps. (*Courtesy of McGraw-Hill*)

Symptoms	Probable cause
Relief valve leaks slowly	Bad relief valve
Relief valve blows off periodically	High water temperature High pressure in tank Bad relief valve
No hot water	Electrical power is off Elements are bad Defective thermostat Inlet valve is closed
Water not hot enough	An element is bad Bad thermostat Thermostat needs adjusting
Water too hot	Thermostat needs adjusting Controls are defective
Water leaks from tank	Hole in tank Rusted-out fitting in tank

FIGURE 12.20 ■ Troubleshooting electric water heaters. (*Courtesy of McGraw-Hill*)

Symptoms	Probable cause
Relief valve leaks slowly	Bad relief valve
Relief valve blows off periodically	High water temperature High pressure in tank Bad relief valve
No hot water	Out of gas Pilot light is out Bad thermostat Control valve is off Gas valve closed
Water not hot enough	Bad thermostat Thermostat needs adjusting
Water too hot	Thermostat needs adjusting Controls are defective Burner will not shut off
Water leaks from tank	Hole in tank Rusted-out fitting in tank

FIGURE 12.21 ■ Troubleshooting gas water heaters. (*Courtesy of McGraw-Hill*)

Symptoms	Probable cause
Wont't drain	Clogged drain Clogged tub waste Clogged trap
Drains slowly	Hair in tub waste Partial drainage blockage
Won't hold water	Tub waste needs adjusting
Won't release water	Tub waste need adjusting
Gurgles as it drains	Partial drainage blockage Partial blockage in the vent
Water drips from spout	Bad faucet washers/cartridge Bad faucet seats
Water comes out spout and shower at the same time	Bad diverter washer Bad diverter seat Bad diverter
Faucet will not shut off	Bad washers or cartridge Bad faucet seats
Poor water pressure	Partially closed valve Not enough water pressure Blockage in the faucet Partially frozen pipe
No water	Closed valve Broken pipe Frozen pipe

FIGURE 12.22 ■ Troubleshooting bathtubs. (*Courtesy of McGraw-Hill*)

Symptoms	Probable cause
Will not flush	No water in tank
	Stoppage in drainage system
Flushes poorly	Clogged flush holes
	Flapper or tank ball is not staying open long enough
	Not enough water in tank
	Partial drain blockage
	Defective handle
	Bad connection between handle and flush valve
	Vent is clogged
Water droplets covering tank	Condensation
Tank fills slowly	Defective ballcock
	Obstructed supply pipe
	Low water pressure
	Partially closed valve
	Partially frozen pipe
Makes unusual noises when flushed	Defective ballcock
Water runs constantly	Bad flapper or tank ball
	Bad ballcock
	Float rod needs adjusting
	Float is filled with water
	Ballcock needs adjusting
	Pitted flush valve
	Undiscovered leak
	Cracked overflow tube
Water seeps from base of toilet	Bad wax ring
	Cracked toilet bowl
Water dripping from tank	Condensation
	Bad tank-to-bowl gasket
	Bad tank-to-bowl bolts
	Cracked tank
	Flush-valve nut is loose
No water comes into the tank	Closed valve
	Defective ballcock
	Frozen pipe
	Broken pipe

FIGURE 12.23 ■ Troubleshooting toilets. (*Courtesy of McGraw-Hill*)

Symptoms	Probable cause
Won't drain	Clogged drain
	Clogged strainer
	Clogged trap
Drains slowly	Hair in strainer
	Partial drainage blockage
Gurgles as it drains	Partial drainage blockage
	Partial blockage in the vent
Water drips from shower head	Bad faucet washers/cartridge
	Bad faucet seats
Faucet will not shut off	Bad washers or cartridge
	Bad faucet seats
Poor water pressure	Partially closed valve
	Not enough water pressure
	Blockage in the faucet
	Partially frozen pipe
No water	Closed valve
	Broken pipe
	Frozen pipe

FIGURE 12.24 ■ Troubleshooting showers. (*Courtesy of McGraw-Hill*)

Symptoms	Probable cause
Faucet drips from spout	Bad washers or cartridge Bad faucet seats
Faucet leaks at base of spout	Bad O ring
Faucet will not shut off	Bad washers or cartridge Bad faucet seats
Poor water pressure	Partially closed valve Clogged aerator Not enough water pressure Blockage in the faucet Partially frozen pipe
No water	Closed valve Broken pipe Frozen pipe
Drains slowly	Hair on pop-up assembly Partial obstruction in drain or trap Pop-up needs to be adjusted
Will not drain	Blocked drain or trap Pop-up is defective
Gurgles as it drains	Partial drainage blockage Partial blockage in the vent
Won't hold water	Pop-up needs adjusting Bad putty seal on drain

FIGURE 12.25 ■ Troubleshooting lavatories. (*Courtesy of McGraw-Hill*)

Symptoms	Probable cause
Faucet drips from spout	Bad washers or cartridge Bad faucet seats
Faucet leaks at base of spout	Bad O ring
Faucet will not shut off	Bad washers or cartridge Bad faucet seats
Poor water pressure	Partially closed valve Clogged aerator Not enough water pressure Blockage in the faucet Partially frozen pipe
No water	Closed valve Broken pipe Frozen pipe
Drains slowly	Partial obstruction in drain or trap
Will not drain	Blocked drain or trap
Gurgles as it drains	Partial drainage blockage Partial blockage in the vent
Won't hold water	Bad basket strainer Bad putty seal on drain

FIGURE 12.26 ▪ Troubleshooting laundry tubs. (*Courtesy of McGraw-Hill*)

Symptoms	Probable cause
Faucet drips from spout	Bad washers or cartridge Bad faucet seats
Faucet leaks at base of spout	Bad O ring
Faucet will not shut off	Bad washers or cartridge Bad faucet seats
Poor water pressure	Partially closed valve Clogged aerator Not enough water pressure Blockage in the faucet Partially frozen pipe
No water	Closed valve Broken pipe Frozen pipe
Drains slowly	Partial obstruction in drain or trap
Will not drain	Blocked drain or trap
Gurgles as it drains	Partial drainage blockage Partial blockage in the vent
Won't hold water	Bad basket strainer Bad putty seal on drain
Spray attachment will not spray	Clogged holes in spray head Kinked spray hose
Spray attachment will not cut off	Bad spray head

FIGURE 12.27 ■ Troubleshooting kitchen sinks. (*Courtesy of McGraw-Hill*)

chapter 13

PLUMBING CODE CONSIDERATIONS

The plumbing code is complex. This chapter is not a replacement for the code, but it will give you a lot of pertinent information that you may use daily in a concise, accessible manner. The majority of the tables provided here were generously provided by the International Code Council, Inc. and the International Plumbing Code 2000. The visual nature of this chapter will allow you to answer many of your code questions by simply reviewing the numerous tables.

In most cases, the tables will speak for themselves. When there may be some confusion, I will provide some insight in the use of a table. For the most part, this is a reference chapter that will not require heavy reading. Consider this your fast track to code facts.

APPROVED MATERIALS

WATER DISTRIBUTION PIPE

MATERIAL	STANDARD
Brass pipe	ASTM B 43
Chlorinated polyvinyl chloride (CPVC) plastic pipe and tubing	ASTM D 2846; ASTM F 441; ASTM F 442; CSA B137.6
Copper or copper-alloy pipe	ASTM B 42; ASTM B 302
Copper or copper-alloy tubing (Type K, WK, L, WL, M or WM)	ASTM B 75; ASTM B 88; ASTM B 251; ASTM B 447
Cross-linked polyethylene (PEX) plastic tubing	ASTM F 877; CSA CAN/CSA-B137.5
Cross-linked polyethylene/ aluminum/cross-linked polyethylene (PEX-AL-PEX) pipe	ASTM F 1281; CSA CAN/CSA-B137.10
Galvanized steel pipe	ASTM A 53
Polybutylene (PB) plastic pipe and tubing	ASTM D 3309; CSA CAN3-B137.8

FIGURE 13.1 ■ Approved materials for water distribution. (*Courtesy of International Code Council, Inc. and the International Plumbing Code 2000*)

WATER SERVICE PIPE

MATERIAL	STANDARD
Acrylonitrile butadiene styrene (ABS) plastic pipe	ASTM D 1527; ASTM D 2282
Asbestos-cement pipe	ASTM C 296
Brass pipe	ASTM B 43
Copper or copper-alloy pipe	ASTM B 42; ASTM B 302
Copper or copper-alloy tubing (Type K, WK, L, WL, M or WM)	ASTM B 75; ASTM B 88; ASTM B 251; ASTM B 447
Chlorinated polyvinyl chloride (CPVC) plastic pipe	ASTM D 2846; ASTM F 441; ASTM F 442; CSA B137.6
Ductile iron water pipe	AWWA C151; AWWA C115
Galvanized steel pipe	ASTM A 53
Polybutylene (PB) plastic pipe and tubing	ASTM D 2662; ASTM D 2666; ASTM D 3309; CSA B137.8
Polyethylene (PE) plastic pipe	ASTM D 2239; CSA CAN/CSA-B137.1
Polyethylene (PE) plastic tubing	ASTM D 2737; CSA B137.1
Cross-linked polyethylene (PEX) plastic tubing	ASTM F 876; ASTM F 877; CSA CAN/CSA-B137.5
Cross-linked polyethylene/ aluminum/cross-linked polyethylene (PEX-AL-PEX) pipe	ASTM F 1281; CSA CAN/CSA B137.10
Polyethylene/aluminum/ polyethylene (PE-AL-PE) pipe	ASTM F 1282; CSA CAN/CSA-B137.9
Polyvinyl chloride (PVC) plastic pipe	ASTM D 1785; ASTM D 2241; ASTM D 2672; CSA CAN/CSA-B137.3

FIGURE 13.2 ■ Approved materials for water service piping. (*Courtesy of International Code Council, Inc. and the International Plumbing Code 2000*)

BUILDING SEWER PIPE

MATERIAL	STANDARD
Acrylonitrile butadiene styrene (ABS) plastic pipe	ASTM D 2661; ASTM D 2751; ASTM F 628
Asbestos-cement pipe	ASTM C 428
Cast-iron pipe	ASTM A 74; ASTM A 888; CISPI 301
Coextruded composite ABS DWV sch 40 IPS pipe (solid)	ASTM F 1488
Coextruded composite ABS DWV sch 40 IPS pipe (cellular core)	ASTM F 1488
Coextruded composite PVC DWV sch 40 IPS pipe (solid)	ASTM F 1488
Coextruded composite PVC DWV sch 40 IPS pipe (cellular core)	ASTM F 1488
Coextruded composite PVC IPS DR - PS DWV, PS140, PS200	ASTM F 1488
Coextruded composite ABS sewer and drain DR - PS in PS35, PS50, PS100, PS140, PS200	ASTM F 1488
Coextruded composite PVC sewer and rain DR - PS in PS35, PS50, PS100, PS140, PS200	ASTM F 1488
Concrete pipe	ASTM C 14; ASTM C 76; CSA A257.1; CSA CAN/CSA A257.2
Copper or copper-alloy tubing (Type K or L)	ASTM B 75; ASTM B 88; ASTM B 251
Polyvinyl chloride (PVC) plastic pipe (Type DWV, SDR26, SDR35, SDR41, PS50 or PS100)	ASTM D 2665; ASTM D 2949; ASTM D 3034; ASTM F 891; CSA B182.2; CSA CAN/CSA-B182.4
Stainless steel drainage systems, Type 316L	ASME/ANSI A112.3.1
Vitrified clay pipe	ASTM C 4; ASTM C 700

FIGURE 13.3 ■ Approved materials for building sewer piping. (*Courtesy of International Code Council, Inc. and the International Plumbing Code 2000*)

UNDERGROUND BUILDING DRAINAGE AND VENT PIPE

MATERIAL	STANDARD
Acrylonitrile butadiene styrene (ABS) plastic pipe	ASTM D 2661; ASTM F 628; CSA B181.1
Asbestos-cement pipe	ASTM C 428
Cast-iron pipe	ASTM A 74; CISPI 301; ASTM A 888
Coextruded composite ABS DWV sch 40 IPS pipe (solid)	ASTM F 1488
Coextruded composite ABS DWV sch 40 IPS pipe (cellular core)	ASTM F 1488
Coextruded composite PVC DWV sch 40 IPS pipe (solid)	ASTM F 1488
Coextruded composite PVC DWV sch 40 IPS pipe (cellular core)	ASTM F 1488
Coextruded composite PVC IPS - DR, PS140, PS200 DWV	ASTM F 1488
Copper or copper alloy tubing (Type K, L, M or DWV)	ASTM B 75; ASTM B 88; ASTM B 251; ASTM B 306
Polyolefin pipe	CSA CAN/CSA-B181.2
Polyvinyl chloride (PVC) plastic pipe (Type DWV)	ASTM D 2665; ASTM D 2949; ASTM F 891; CSA CAN/CSA-B181.2
Stainless steel drainage systems, Type 316L	ASME/ANSI A112.3.1

FIGURE 13.4 ■ Approved materials for underground building drainage and vent pipe. (*Courtesy of International Code Council, Inc. and the International Plumbing Code 2000*)

PIPE FITTINGS

MATERIAL	STANDARD
Acrylonitrile butadiene styrene (ABS) plastic	ASTM D 3311; CSA B181.1; ASTM D 2661
Cast iron	ASME B16.4; ASME B16.12; ASTM A 74; ASTM A 888; CISPI 301
Coextruded composite ABS DWV sch 40 IPS pipe (solid or cellular core)	ASTM D 2661; ASTM D 3311; ASTM F 628
Coextruded composite PVC DWV sch 40 pipe IPS-DR, PS140, PS200 (solid or cellular core)	ASTM D 2665; ASTM D 3311; ASTM F 891
Coextruded composite ABS sewer and drain DR-PS in PS35, PS50, PS100, PS140, PS200	ASTM D 2751
Coextruded composite PVC sewer and drain DR-PS in PS35, PS50, PS100, PS140, PS200	ASTM D 3034
Copper or copper alloy	ASME B16.15; ASME B16.18; ASME B16.22; ASME B16.23; ASME B16.26; ASME B16.29; ASME B16.32
Glass	ASTM C 1053
Gray iron and ductile iron	AWWA C110
Malleable iron	ASME B16.3
Polyvinyl chloride (PVC) plastic	ASTM D 3311; ASTM D 2665
Stainless steel drainage systems, Types 304 and 316L	ASME/ANSI A112.3.1
Steel	ASME B16.9; ASME B16.11; ASME B16.28

FIGURE 13.5 ■ Approved materials for above-ground drainage and vent pipe. (*Courtesy of International Code Council, Inc. and the International Plumbing Code 2000*)

ABOVE-GROUND DRAINAGE AND VENT PIPE

MATERIAL	STANDARD
Acrylonitrile butadiene styrene (ABS) plastic pipe	ASTM D 2661; ASTM F 628; CSA B181.1
Brass pipe	ASTM B 43
Cast-iron pipe	ASTM A 74; CISPI 301; ASTM A 888
Coextruded composite ABS DWV sch 40 IPS pipe (solid)	ASTM F 1488
Coextruded composite ABS DWV sch 40 IPS pipe (cellular core)	ASTM F 1488
Coextruded composite PVC DWV sch 40 IPS pipe (solid)	ASTM F 1488
Coextruded composite PVC DWV sch 40 IPS pipe (cellular core)	ASTM F 1488
Coextruded composite PVC IPS - DR, PS140, PS200 DWV	ASTM F 1488
Copper or copper-alloy pipe	ASTM B 42; ASTM B 302
Copper or copper-alloy tubing (Type K, L, M or DWV)	ASTM B 75; ASTM B 88; ASTM B 251; ASTM B 306
Galvanized steel pipe	ASTM A 53
Glass pipe	ASTM C 1053
Polyolefin pipe	CSA CAN/CSA-B181.2
Polyvinyl chloride (PVC) plastic pipe (Type DWV)	ASTM D 2665; ASTM D 2949; ASTM F 891; CSA CAN/CSA-B181.2; ASTM F 1488
Stainless steel drainage systems, types 304 and 316L	ASME/ANSI A112.3.1

FIGURE 13.6 ▪ Approved materials for pipe fittings. (*Courtesy of International Code Council, Inc. and the International Plumbing Code 2000*)

SIZE OF PIPE IDENTIFICATION

PIPE DIAMETER (Inches)	LENGTH OF BACKGROUND COLOR FIELD (Inches)	SIZE OF LETTERS (Inches)
$3/4$ to $1^1/4$	8	0.5
$1^1/2$ to 2	8	0.75
$2^1/2$ to 6	12	1.25
8 to 10	24	2.5
over 10	32	3.5

For SI: 1 inch = 25.4 mm.

FIGURE 13.7 ▪ Requirements of pipe identification. (*Courtesy of International Code Council, Inc. and the International Plumbing Code 2000*)

Caulking Ferrules

Pipe size (inches)	Inside diameter (inches)	Length (inches)	Minimum weight each Lb.	Oz.
2	2-1/4	4-1/2	1	0
3	3-1/4	4-1/2	1	12
4	4-1/4	4-1/2	2	8

Caulking Ferrules (Metric)

Pipe size (mm)	Inside diameter (mm)	Length (mm)	Minimum weight each (kg)
50	57	114	0.454
80	83	114	0.790
100	108	114	1.132

Soldering Bushings

Pipe size (inches)	Minimum weight each Lb.	Oz.	Pipe size (inches)	Minimum weight each Lb.	Oz.
1-1/4	0	6	2-1/2	1	6
1-1/2	0	8	3	2	0
2	0	14	4	3	8

Soldering Bushings (Metric)

Pipe size (mm)	Minimum weight each (kg)	Pipe size (mm)	Minimum weight each (kg)
32	0.168	65	0.622
40	0.224	80	0.908
50	0.392	100	1.586

FIGURE 13.8 ▪ Requirements for ferrules and bushings. (*Courtesy of International Code Council, Inc. and the International Plumbing Code 2000*)

MINIMUM PLUMBING FACILITIES

MINIMUM NUMBER OF PLUMBING FACILITIES[a]

	OCCUPANCY	WATER CLOSETS (Urinals, see Section 419.2) Male	Female	LAVATORIES	BATHTUBS/ SHOWERS	DRINKING FOUNTAINS (see Section 410.1)	OTHERS
A S S E M B L Y	Nightclubs	1 per 40	1 per 40	1 per 75	—	1 per 500	1 service sink
	Restaurants	1 per 75	1 per 75	1 per 200	—	1 per 500	1 service sink
	Theaters, halls, museums, etc.	1 per 125	1 per 65	1 per 200	—	1 per 500	1 service sink
	Coliseums, arenas (less than 3,000 seats)	1 per 75	1 per 40	1 per 150	—	1 per 1,000	1 service sink
	Coliseums, arenas (3,000 seats or greater)	1 per 120	1 per 60	Male 1 per 200 Female 1 per 150	—	1 per 1,000	1 service sink
	Churches[b]	1 per 150	1 per 75	1 per 200	—	1 per 1,000	1 service sink
	Stadiums (less than 3,000 seats), pools, etc.	1 per 100	1 per 50	1 per 150	—	1 per 1,000	1 service sink
	Stadiums (3,000 seats or greater)	1 per 150	1 per 75	Male 1 per 200 Female 1 per 150	—	1 per 1,000	1 service sink
I N S T I T U T I O N A L	Business (see Sections 403.2, 403.4 and 403.5)	1 per 50		1 per 80	—	1 per 100	1 service sink
	Educational	1 per 50		1 per 50	—	1 per 100	1 service sink
	Factory and industrial	1 per 100		1 per 100	(see Section 411)	1 per 400	1 service sink
	Passenger terminals and transportation facilities	1 per 500		1 per 750	—	1 per 1,000	1 service sink
	Residential care	1 per 10		1 per 10	1 per 8	1 per 100	1 service sink
	Hospitals, ambulatory nursing home patients[c]	1 per room[d]		1 per room[d]	1 per 15	1 per 100	1 service sink per floor
	Day nurseries, sanitariums, nonambulatory nursing home patients, etc.[c]	1 per 15		1 per 15	1 per 15[e]	1 per 100	1 service sink
	Employees, other than residential care[c]	1 per 25		1 per 35	—	1 per 100	—
	Visitors, other than residential care	1 per 75		1 per 100	—	1 per 500	—
	Prisons[c]	1 per cell		1 per cell	1 per 15	1 per 100	1 service sink
	Asylums, reformatories, etc.[c]	1 per 15		1 per 15	1 per 15	1 per 100	1 service sink
	Mercantile (see Sections 403.2, 403.4 and 403.5)	1 per 500		1 per 750	—	1 per 1,000	1 service sink
R E S I D E N T I A L	Hotels, motels	1 per guestroom		1 per guestroom	1 per guestroom	—	1 service sink
	Lodges	1 per 10		1 per 10	1 per 8	1 per 100	1 service sink
	Multiple family	1 per dwelling unit		1 per dwelling unit	1 per dwelling unit	—	1 kitchen sink per dwelling unit; 1 automatic clothes washer connection per 20 dwelling units
	Dormitories	1 per 10		1 per 10	1 per 8	1 per 100	1 service sink
	One- and two-family dwellings	1 per dwelling unit		1 per dwelling unit	1 per dwelling unit	—	1 kitchen sink per dwelling unit; 1 automatic clothes washer connection per dwelling unit[f]
	Storage (see Sections 403.2 and 403.4)	1 per 100		1 per 100	(see Section 411)	1 per 1,000	1 service sink

a. The fixtures shown are based on one fixture being the minimum required for the number of persons indicated or any fraction of the number of persons indicated. The number of occupants shall be determined by the *International Building Code*.

b. Fixtures located in adjacent buildings under the ownership or control of the church shall be made available during periods the church is occupied.

c. Toilet facilities for employees shall be separate from facilities for inmates or patients.

d. A single-occupant toilet room with one water closet and one lavatory serving not more than two adjacent patient rooms shall be permitted where such room is provided with direct access from each patient room and with provisions for privacy.

e. For day nurseries, a maximum of one bathtub shall be required.

f. For attached one- and two-family dwellings, one automatic clothes washer connection shall be required per 20 dwelling units.

FIGURE 13.9 ■ Minimum plumbing facilities. (*Courtesy of International Code Council, Inc. and the International Plumbing Code 2000*)

MINIMUM SIZES OF FIXTURE WATER SUPPLY PIPES

FIXTURE	MINIMUM PIPE SIZE (inch)
Bathtubs (60″ × 32″ and smaller)[a]	$1/2$
Bathtubs (larger than 60″ × 32″)	$1/2$
Bidet	$3/8$
Combination sink and tray	$1/2$
Dishwasher, domestic[a]	$1/2$
Drinking fountain	$3/8$
Hose bibbs	$1/2$
Kitchen sink[a]	$1/2$
Laundry, 1, 2 or 3 compartments[a]	$1/2$
Lavatory	$3/8$
Shower, single head[a]	$1/2$
Sinks, flushing rim	$3/4$
Sinks, service	$1/2$
Urinal, flush tank	$1/2$
Urinal, flush valve	$3/4$
Wall hydrant	$1/2$
Water closet, flush tank	$3/8$
Water closet, flush valve	1
Water closet, flushometer tank	$3/8$
Water closet, one piece[a]	$1/2$

For SI: 1 inch = 25.4 mm, 1 foot = 304.8 mm, 1 psi = 6.895 kPa.

a. Where the developed length of the distribution line is 60 feet or less, and the available pressure at the meter is a minimum of 35 psi, the minimum size of an individual distribution line supplied from a manifold and installed as part of a parallel water distribution system shall be one nominal tube size smaller than the sizes indicated.

FIGURE 13.10 ■ Minimum number of plumbing facilities. (*Courtesy of International Code Council, Inc. and the International Plumbing Code 2000*)

FIXTURE SUPPLIES

**MAXIMUM FLOW RATES AND CONSUMPTION
FOR PLUMBING FIXTURES AND FIXTURE FITTINGS**

PLUMBING FIXTURE OR FIXTURE FITTING	MAXIMUM FLOW RATE OR QUANTITY[b]
Water closet	1.6 gallons per flushing cycle
Urinal	1.0 gallon per flushing cycle
Shower head[a]	2.5 gpm at 60 psi
Lavatory, private	2.2 gpm at 60 psi
Lavatory (other than metering), public	0.5 gpm at 60 psi
Lavatory, public (metering)	0.25 gallon per metering cycle
Sink faucet	2.2 gpm at 60 psi

For SI: 1 gallon = 3.785 L, 1 gallon per minute = 3.785 L/m, 1 psi = 6.895 kPa.

a. A hand-held shower spray is a shower head.

b. Consumption tolerances shall be determined from referenced standards.

FIGURE 13.11 ■ Maximum flow rates and consumption for plumbing fixtures and fixture fittings. (*Courtesy of International Code Council, Inc. and the International Plumbing Code 2000*)

MANIFOLD SIZING

NOMINAL SIZE INTERNAL DIAMETER (inches)	MAXIMUM DEMAND (gpm)	
	Velocity at 4 feet per second	Velocity at 8 feet per second
$1/2$	2	5
$3/4$	6	11
1	10	20
$1 1/4$	15	31
$1 1/2$	22	44

For SI: 1 inch = 25.4 mm, 1 gallon per minute = 3.785 L/m, 1 foot per second = 0.305 m/s.

FIGURE 13.12 ■ Maximum flow rates and consumption for plumbing fixtures and fixture fittings. (*Courtesy of International Code Council, Inc. and the International Plumbing Code 2000*)

**WATER DISTRIBUTION SYSTEM DESIGN CRITERIA
REQUIRED CAPACITIES AT FIXTURE SUPPLY PIPE OUTLETS**

FIXTURE SUPPLY OUTLET SERVING	FLOW RATE[a] (gpm)	FLOW PRESSURE (psi)
Bathtub	4	8
Bidet	2	4
Combination fixture	4	8
Dishwasher, residential	2.75	8
Drinking fountain	0.75	8
Laundry tray	4	8
Lavatory	2	8
Shower	3	8
Shower, temperature controlled	3	20
Sillcock, hose bibb	5	8
Sink, residential	2.5	8
Sink, service	3	8
Urinal, valve	15	15
Water closet, blow out, flushometer valve	35	25
Water closet, flushometer tank	1.6	15
Water closet, siphonic, flushometer valve	25	15
Water closet, tank, close coupled	3	8
Water closet, tank, one piece	6	20

For SI: 1 psi = 6.895 kPa, 1 gallon per minute (gpm) = 3.785 L/m.

a. For additional requirements for flow rates and quantities, see Section 604.4.

FIGURE 13.13 ■ Water distribution system design criteria required capacities at fixture supply pipe outlets. (*Courtesy of International Code Council, Inc. and the International Plumbing Code 2000*)

HORIZONTAL FIXTURE BRANCHES AND STACKS[a]

DIAMETER OF PIPE (inches)	MAXIMUM NUMBER OF DRAINAGE FIXTURE UNITS (dfu)			
	Total for a horizontal branch	Total discharge into one branch interval	Stacks[b]	
			Total for stack of three branch intervals or less	Total for stack greater than three branch intervals
1½	3	2	4	8
2	6	6	10	24
2½	12	9	20	42
3	20	20	48	72
4	160	90	240	500
5	360	200	540	1,100
6	620	350	960	1,900
8	1,400	600	2,200	3,600
10	2,500	1,000	3,800	5,600
12	2,900	1,500	6,000	8,400
15	7,000	Footnote c	Footnote c	Footnote c

For SI: 1 inch = 25.4 mm.

a. Does not include branches of the building drain. Refer to Table 710.1(1).

b. Stacks shall be sized based on the total accumulated connected load at each story or branch interval. As the total accumulated connected load decreases, stacks are permitted to be reduced in size. Stack diameters shall not be reduced to less than one-half of the diameter of the largest stack size required.

c. Sizing load based on design criteria.

FIGURE 13.14 ■ Drainage fixture units allowed on horizontal fixture branches and stacks. *(Courtesy of International Code Council, Inc. and the International Plumbing Code 2000)*

DWV DATA

BUILDING DRAINS AND SEWERS

DIAMETER OF PIPE (inches)	MAXIMUM NUMBER OF DRAINAGE FIXTURE UNITS CONNECTED TO ANY PORTION OF THE BUILDING DRAIN OR THE BUILDING SEWER, INCLUDING BRANCHES OF THE BUILDING DRAIN[a]			
	Slope per foot			
	1/16 inch	1/8 inch	1/4 inch	1/2 inch
1¼	—	—	1	1
1½	—	—	3	3
2	—	—	21	26
2½	—	—	24	31
3	—	36	42	50
4	—	180	216	250
5	—	390	480	575
6	—	700	840	1,000
8	1,400	1,600	1,920	2,300
10	2,500	2,900	3,500	4,200
12	3,900	4,600	5,600	6,700
15	7,000	8,300	10,000	12,000

For SI: 1 inch = 25.4 mm, 1 inch per foot = 0.0833 mm/m.

a. The minimum size of any building drain serving a water closet shall be 3 inches.

FIGURE 13.15 ▪ Drainage fixture units allowed for building drains and sewers. *(Courtesy of International Code Council, Inc. and the International Plumbing Code 2000)*

Maximum Unit Loading and Maximum Length of Drainage and Vent Piping

Size of Pipe, Inches (mm)	1-1/4 (32)	1-1/2 (40)	2 (50)	2-1/2 (65)	3 (80)	4 (100)	5 (125)	6 (150)	8 (200)	10 (250)	12 (300)
Maximum Units											
Drainage Piping[1]											
Vertical	1	2[2]	16[3]	32[3]	48[4]	256	600	1380	3600	5600	8400
Horizontal	1	1	8[3]	14[3]	35[4]	216[5]	428[5]	720[5]	2640[5]	4680[5]	8200[5]
Maximum Length											
Drainage Piping											
Vertical, feet	45	65	85	148	212	300	390	510	750		
(m)	(14)	(20)	(26)	(45)	(65)	(91)	(119)	(155)	(228)		
Horizontal (Unlimited)											
Vent Piping (See note)											
Horizontal and Vertical											
Maximum Units	1	8[3]	24	48	84	256	600	1380	3600		
Maximum Lengths, feet	45	60	120	180	212	300	390	510	750		
(m)	(14)	(18)	(37)	(55)	(65)	(91)	(119)	(155)	(228)		

1 Excluding trap arm.
2 Except sinks, urinals and dishwashers.
3 Except six-unit traps or water closets.
4 Only four (4) water closets or six-unit traps allowed on any vertical pipe or stack; and not to exceed three (3) water closets or six-unit traps on any horizontal branch or drain.
5 Based on one-fourth (1/4) inch per foot (20.9 mm/m) slope. For one-eighth (1/8) inch per foot (10.4 mm/m) slope, multiply horizontal fixture units by a factor of 0.8.

Note: The diameter of an individual vent shall not be less than one and one-fourth (1-1/4) inches (31.8 mm) nor less than one-half (1/2) the diameter of the drain to which it is connected. Fixture unit load values for drainage and vent piping shall be computed from Tables 7-3 and 7-4. Not to exceed one-third (1/3) of the total permitted length of any vent may be installed in a horizontal position. When vents are increased one (1) pipe size for their entire length, the maximum length limitations specified in this table do not apply.

FIGURE 13.16 ■ Maximum unit loading and maximum length of drainage and vent piping. (Courtesy of International Code Council, Inc. and the International Plumbing Code 2000)

SIZE OF COMBINATION DRAIN AND VENT PIPE

DIAMETER PIPE (inches)	MAXIMUM NUMBER OF DRAINAGE FIXTURE UNITS (dfu)	
	Connecting to a horizontal branch or stack	Connecting to a building drain or building subdrain
2	3	4
$2^1/_2$	6	26
3	12	31
4	20	50
5	160	250
6	360	575

For SI: 1 inch = 25.4 mm.

FIGURE 13.17 ▪ Size of combination drain and vent pipe. (*Courtesy of International Code Council, Inc. and the International Plumbing Code 2000*)

SLOPE OF HORIZONTAL DRAINAGE PIPE

SIZE (inches)	MINIMUM SLOPE (inch per foot)
$2^1/_2$ or less	$^1/_4$
3 to 6	$^1/_8$
8 or larger	$^1/_{16}$

For SI: 1 inch = 25.4 mm, 1 inch per foot = 0.0833 mm/m.

FIGURE 13.18 ▪ Slope of horizontal drainage pipe. (*Courtesy of International Code Council, Inc. and the International Plumbing Code 2000*)

DRAINAGE FIXTURE UNITS FOR FIXTURE DRAINS OR TRAPS

FIXTURE DRAIN OR TRAP SIZE (inches)	DRAINAGE FIXTURE UNIT VALUE
$1^1/_4$	1
$1^1/_2$	2
2	3
$2^1/_2$	4
3	5
4	6

For SI: 1 inch = 25.4 mm.

FIGURE 13.19 ▪ Drainage fixture units for fixture drains or traps. (*Courtesy of International Code Council, Inc. and the International Plumbing Code 2000*)

MINIMUM CAPACITY OF SEWAGE PUMP OR SEWAGE EJECTOR

DIAMETER OF THE DISCHARGE PIPE (Inches)	CAPACITY OF PUMP OR EJECTOR (gpm)
2	21
$2^1/_2$	30
3	46

For SI: 1 inch = 25.4 mm, 1 gpm = 3.785 L/m.

FIGURE 13.20 ■ Minimum capacity of sewage pump or sewage ejector. (*Courtesy of International Code Council, Inc. and the International Plumbing Code 2000*)

DRAINAGE FIXTURE UNITS FOR FIXTURES AND GROUPS

FIXTURE TYPE	DRAINAGE FIXTURE UNIT VALUE AS LOAD FACTORS	MINIMUM SIZE OF TRAP (Inches)
Automatic clothes washers, commercial[a]	3	2
Automatic clothes washers, residential	2	2
Bathroom group as defined in Section 202 (1.6 gpf water closet)[f]	5	—
Bathroom group as defined in Section 202 (water closet flushing greater than 1.6 gpf)[f]	6	—
Bathtub[b] (with or without overhead shower or whirlpool attachments)	2	$1^1/_2$
Bidet	1	$1^1/_4$
Combination sink and tray	2	$1^1/_2$
Dental lavatory	1	$1^1/_4$
Dental unit or cuspidor	1	$1^1/_4$
Dishwashing machine,[c] domestic	2	$1^1/_2$
Drinking fountain	$1/_2$	$1^1/_4$
Emergency floor drain	0	2
Floor drains	2	2
Kitchen sink, domestic	2	$1^1/_2$
Kitchen sink, domestic with food waste grinder and/or dishwasher	2	$1^1/_2$
Laundry tray (1 or 2 compartments)	2	$1^1/_2$
Lavatory	1	$1^1/_4$
Shower	2	$1^1/_2$
Sink	2	$1^1/_2$
Urinal	4	Footnote d
Urinal, 1 gallon per flush or less	2[e]	Footnote d
Wash sink (circular or multiple) each set of faucets	2	$1^1/_2$
Water closet, flushometer tank, public or private	4[e]	Footnote d
Water closet, private (1.6 gpf)	3[e]	Footnote d
Water closet, private (flushing greater than 1.6 gpf)	4[e]	Footnote d
Water closet, public (1.6 gpf)	4[e]	Footnote d
Water closet, public (flushing greater than 1.6 gpf)	6[e]	Footnote d

For SI: 1 inch = 25 4 mm, 1 gallon = 3.785 L

a. For traps larger than 3 inches, use Table 709.2.
b. A showerhead over a bathtub or whirlpool bathtub attachments does not increase the drainage fixture unit value.
c. See Sections 709.2 through 709.4 for methods of computing unit value of fixtures not listed in Table 709.1 or for rating of devices with intermittent flows.
d. Trap size shall be consistent with the fixture outlet size.
e. For the purpose of computing loads on building drains and sewers, water closets or urinals shall not be rated at a lower drainage fixture unit unless the lower values are confirmed by testing.
f. For fixtures added to a dwelling unit bathroom group, add the DFU value of those additional fixtures to the bathroom group fixture count.

FIGURE 13.21 ■ Drainage fixture units. (*Courtesy of International Code Council, Inc. and the International Plumbing Code 2000*)

MAXIMUM DISTANCE OF FIXTURE TRAP FROM VENT

SIZE OF TRAP (inches)	SIZE OF FIXTURE DRAIN (inches)	SLOPE (inch per foot)	DISTANCE FROM TRAP (feet)
$1^1/_4$	$1^1/_4$	$^1/_4$	$3^1/_2$
$1^1/_4$	$1^1/_2$	$^1/_4$	5
$1^1/_2$	$1^1/_2$	$^1/_4$	5
$1^1/_2$	2	$^1/_4$	6
2	2	$^1/_4$	6
3	3	$^1/_8$	10
4	4	$^1/_8$	12

For SI: 1 inch = 25.4 mm, 1 foot = 304.8 mm, 1 inch per foot = 0.0833 mm/m.

FIGURE 13.22 ■ Maximum distance of fixture trap from vent.(*Courtesy of International Code Council, Inc. and the International Plumbing Code 2000*)

COMMON VENT SIZES

PIPE SIZE (inches)	MAXIMUM DISCHARGE FROM UPPER FIXTURE DRAIN (dfu)
$1^1/_2$	1
2	4
$2^1/_2$ to 3	6

For SI: 1 inch = 25.4 mm.

FIGURE 13.23 ■ Common vent sizes. (*Courtesy of International Code Council, Inc. and the International Plumbing Code 2000*)

SIZE AND LENGTH OF SUMP VENTS

DISCHARGE CAPACITY OF PUMP (gpm)	MAXIMUM DEVELOPED LENGTH OF VENT (feet)[a]					
	Diameter of vent (inches)					
	1¼	1½	2	2½	3	4
10	No limit[b]	No limit	No limit	No limit	No limit	No limit
20	270	No limit	No limit	No limit	No limit	No limit
40	72	160	No limit	No limit	No limit	No limit
60	31	75	270	No limit	No limit	No limit
80	16	41	150	380	No limit	No limit
100	10[c]	25	97	250	No limit	No limit
150	Not permitted	10[c]	44	110	370	No limit
200	Not permitted	Not permitted	20	60	210	No limit
250	Not permitted	Not permitted	10	36	132	No limit
300	Not permitted	Not permitted	10[c]	22	88	380
400	Not permitted	Not permitted	Not permitted	10[c]	44	210
500	Not permitted	Not permitted	Not permitted	Not permitted	24	130

For SI: 1 inch = 25.4 mm, 1 foot = 304.8 mm, 1 gallons per minute = 3.785 L/m.

a. Developed length plus an appropriate allowance for entrance losses and friction due to fittings, changes in direction and diameter. Suggested allowances shall be obtained from NBS Monograph 31 or other approved sources. An allowance of 50 percent of the developed length shall be assumed if a more precise value is not available.

b. Actual values greater than 500 feet.

c. Less than 10 feet.

FIGURE 13.24 ■ Size and length of sump vents. *(Courtesy of International Code Council, Inc. and the International Plumbing Code 2000)*

MINIMUM DIAMETER AND MAXIMUM LENGTH OF INDIVIDUAL BRANCH FIXTURE VENTS AND INDIVIDUAL FIXTURE HEADER VENTS FOR SMOOTH PIPES

DIAMETER OF VENT PIPE (inches)	INDIVIDUAL VENT AIRFLOW RATE (cubic feet per minute)																			
	Maximum developed length of vent (feet)																			
	1	2	3	4	5	6	7	8	9	10	11	12	13	14	15	16	17	18	19	20
$1/_2$	95	25	13	8	5	4	3	2	1	1	1	1	1	1	1	1	1	1	1	1
$3/_4$	100	88	47	30	20	15	10	9	7	6	5	4	3	3	3	2	2	2	2	1
1	—	—	100	94	65	48	37	29	24	20	17	14	12	11	9	8	7	7	6	6
$1^1/_4$	—	—	—	—	—	—	—	100	87	73	62	53	46	40	36	32	29	26	23	21
$1^1/_2$	—	—	—	—	—	—	—	—	—	—	—	100	96	84	75	67	60	54	49	45
2	—	—	—	—	—	—	—	—	—	—	—	—	—	—	—	—	—	—	—	100

For SI: 1 inch = 25.4 mm, 1 foot = 304.8 mm, 1 cfm = 0.4719 L/s.

FIGURE 13.25 ▪ Minimum diameter and maximum length of individual branch fixture vents and individual fixture header vents for smooth pipes. (*Courtesy of International Code Council, Inc. and the International Plumbing Code 2000*)

SIZE AND DEVELOPED LENGTH OF STACK VENTS AND VENT STACKS

DIAMETER OF SOIL OR WASTE STACK (inches)	TOTAL FIXTURE UNITS BEING VENTED (dfu)	MAXIMUM DEVELOPED LENGTH OF VENT (feet) DIAMETER OF VENT (inches)										
		1¼	1½	2	2½	3	4	5	6	8	10	12
1¼	2	30										
1½	8	50	150									
1½	10	30	100									
2	12	30	75	200								
2	20	26	50	150								
2½	42		30	100	300							
3	10		42	150	360	1,040						
3	21		32	110	270	810						
3	53		27	94	230	680						
3	102		25	86	210	620						
4	43			35	85	250	980					
4	140			27	65	200	750					
4	320			23	55	170	640					
4	540			21	50	150	580					
5	190				28	82	320	990				
5	490				21	63	250	760				
5	940				18	53	210	670				
5	1,400				16	49	190	590				
6	500					33	130	400	1,000			
6	1,100					26	100	310	780			
6	2,000					22	84	260	660			
6	2,900					20	77	240	600			
8	1,800						31	95	240	940		
8	3,400						24	73	190	720		
8	5,600						20	62	160	610		
8	7,600						18	56	140	560		
10	4,000							31	78	310	960	
10	7,200							24	60	240	740	
10	11,000							20	51	200	630	
10	15,000							18	46	180	570	
12	7,300								31	120	380	940
12	13,000								24	94	300	720
12	20,000								20	79	250	610
12	26,000								18	72	230	500
15	15,000									40	130	310
15	25,000									31	96	240
15	38,000									26	81	200
15	50,000									24	74	180

FIGURE 13.26 ▪ Size and developed length of stack vents and vent stacks. (*Courtesy of International Code Council, Inc. and the International Plumbing Code 2000*)

SIZE OF DRAIN PIPES FOR WATER TANKS

TANK CAPACITY (gallons)	DRAIN PIPE (inches)
Up to 750	1
751 to 1,500	1½
1,501 to 3,000	2
3,001 to 5,000	2½
5,001 to 7,500	3
Over 7,500	4

For SI: 1 inch = 25.4 mm, 1 gallon = 3.785 L.

FIGURE 13.27 ▪ Size of drain pipes for water tanks. (*Courtesy of International Code Council, Inc. and the International Plumbing Code 2000*)

HANGER SPACING

PIPING MATERIAL	MAXIMUM HORIZONTAL SPACING (feet)	MAXIMUM VERTICAL SPACING (feet)
ABS pipe	4	10[b]
Aluminum tubing	10	15
Brass pipe	10	10
Cast-iron pipe[a]	5	15
Copper or copper-alloy pipe	12	10
Copper or copper-alloy tubing, $1^1/_4$-inch diameter and smaller	6	10
Copper or copper-alloy tubing, $1^1/_2$-inch diameter and larger	10	10
Cross-linked polyethylene (PEX) pipe	2.67 (32 inches)	10[b]
Cross-linked polyethylene/ aluminum/crosslinked polyethylene (PEX-AL-PEX) pipe	$2^2/_3$ (32 inches)	4
CPVC pipe or tubing, 1 inch or smaller	3	10[b]
CPVC pipe or tubing, $1^1/_4$ inches or larger	4	10[b]
Steel pipe	12	15
Lead pipe	Continuous	4
PB pipe or tubing	2.67 (32 inches)	4
Polyethylene/aluminum/polyethylene (PE-AL-PE) pipe	2.67 (32 inches)	4
PVC pipe	4	10[b]
Stainless steel drainage systems	10	10[b]

For SI: 1 inch = 25.4 mm, 1 foot = 304.8 mm.

a. The maximum horizontal spacing of cast-iron pipe hangers shall be increased to 10 feet where 10-foot lengths of pipe are installed.

b. Midstory guide for sizes 2 inches and smaller.

FIGURE 13.28 ▪ Hanger spacing. (*Courtesy of International Code Council, Inc. and the International Plumbing Code 2000*)

SIZES FOR OVERFLOW PIPES FOR WATER SUPPLY TANKS

MAXIMUM CAPACITY OF WATER SUPPLY LINE TO TANK (gpm)	DIAMETER OF OVERFLOW PIPE (inches)
0 - 50	2
50 - 150	$2^1/_2$
150 - 200	3
200 - 400	4
400 - 700	5
700 - 1,000	6
Over 1,000	8

For SI: 1 inch = 25.4 mm, 1 gallon per minute = 3.785 L/m.

FIGURE 13.29 ■ Sizes for overflow pipes for water supply tanks. (*Courtesy of International Code Council, Inc. and the International Plumbing Code 2000*)

Materials	Type of Joints	Horizontal	Vertical
Cast Iron Hub and Spigot	Lead and Oakum	5 feet (1524 mm), except may be 10 feet (3048 mm) where 10 foot (3048 mm) lengths are installed [1, 2, 3]	Base and each floor not to exceed 15 feet (4572 mm)
	Compression Gasket	Every other joint, unless over 4 feet (1219 mm), then support each joint [1,2,3]	Base and each floor not to exceed 15 feet (4572 mm)
Cast Iron Hubless	Shielded Coupling	Every other joint, unless over 4 feet (1249 mm), then support each joint [1,2,3,4]	Base and each floor not to exceed 15 feet (4572 mm)
Copper Tube and Pipe	Soldered, Brazed or Welded	1-1/2 inch (40 mm) and smaller, 6 feet (1829 mm), 2 inch (50 mm) and larger, 10 feet (3048 mm)	Each floor, not to exceed 10 feet (3048 mm) [5]
Steel and Brass Pipe for Water or DWV	Threaded or Welded	3/4 inch (20 mm) and smaller, 10 feet (3048 mm), 1 inch (25 mm) and larger, 12 feet (3658 mm)	Every other floor, not to exceed 25 feet (7620 mm) [5]
Steel, Brass and Tinned Copper Pipe for Gas	Threaded or Welded	1/2 inch (15 mm), 6 feet (1829 mm) 3/4 (20 mm) and 1 inch (25.4 mm), 8 feet (2438 mm), 1-1/4 inch (32 mm) and larger, 10 feet (3048 mm)	1/2 inch (12.7 mm), 6 feet (1829 mm), 3/4 (19 mm) and 1 inch (25.4 mm), 8 feet (2438 mm), 1-1/4 inch (32 mm) and larger, every floor level
Schedule 40 PVC and ABS DWV	Solvent Cemented	All sizes, 4 feet (1219 mm). Allow for expansion every 30 feet (9144 mm) [3, 6]	Base and each floor. Provide mid-story guides. Provide for expansion every 30 feet (9144 mm) [6]
CPVC	Solvent Cemented	1 inch (25 mm) and smaller, 3 feet (914 mm), 1-1/4 inch (32 mm) and larger, 4 feet (1219 mm)	Base and each floor. Provide mid-story guides [6]
Lead	Wiped or Burned	Continuous support	Not to exceed 4 feet (1219 mm)
Copper	Mechanical	In accordance with standards acceptable to the Administrative Authority	
Steel & Brass	Mechanical	In accordance with standards acceptable to the Administrative Authority	
PEX	Metal Insert and Metal Compression	32 inches (800 mm)	Base and each floor. Provide midstory guides

[1] Support adjacent to joint, not to exceed eighteen (18) inches (457 mm).
[2] Brace at not more than forty (40) foot (12192 mm) intervals to prevent horizontal movement.
[3] Support at each horizontal branch connection.
[4] Hangers shall not be placed on the coupling.
[5] Vertical water lines may be supported in accordance with recognized engineering principles with regard to expansion and contraction, when first approved by the Administrative Authority.
[6] See the appropriate IAPMO Installation Standard for expansion and other special requirements.

FIGURE 13.30 ■ Horizontal and vertical use of materials and joints. (*Courtesy of International Code Council, Inc. and the International Plumbing Code 2000*)

MINIMUM REQUIRED AIR GAPS

FIXTURE	MINIMUM AIR GAP	
	Away from a wall[a] (inches)	Close to a wall (inches)
Lavatories and other fixtures with effective opening not greater than $1/_2$ inch in diameter	1	$1^1/_2$
Sink, laundry trays, gooseneck back faucets and other fixtures with effective openings not greater than $3/_4$ inch in diameter	1.5	2.5
Over-rim bath fillers and other fixtures with effective openings not greater than 1 inch in diameter	2	3
Drinking water fountains, single orifice not greater than $7/_{16}$ inch in diameter or multiple orifices with a total area of 0.150 square inch (area of circle $7/_{16}$ inch in diameter)	1	$1^1/_2$
Effective openings greater than 1 inch	Two times the diameter of the effective opening	Three times the diameter of the effective opening

For SI: 1 inch = 25.4 mm.

a. Applicable where walls or obstructions are spaced from the nearest inside edge of the spout opening a distance greater than three times the diameter of the effective opening for a single wall, or a distance greater than four times the diameter of the effective opening for two intersecting walls.

FIGURE 13.31 ■ Minimum required air gaps. (*Courtesy of International Code Council, Inc. and the International Plumbing Code 2000*)

AIRGAPS AND AIR CHAMBERS

Minimum Airgaps for Water Distribution[4]		
Fixtures	When not affected by side walls[1] Inches (mm)	When affected by side walls[2] Inches (mm)
Effective openings[3] not greater than one-half (1/2) inch (12.7 mm) in diameter	1 (25.4)	1-1/2 (38)
Effective openings[3] not greater than three-quarters (3/4) inch (20 mm) in diameter	1-1/2 (38)	2-1/4 (57)
Effective openings[3] not greater than one (1) inch (25 mm) in diameter	2 (51)	3 (76)
Effective openings[3] greater than one (1) inch (25 mm) in diameter	Two (2) times diameter of effective opening	Three (3) times diameter of effective opening

[1] Side walls, ribs or similar obstructions do not affect airgaps when spaced from the inside edge of the spout opening a distance greater than three times the diameter of the effective opening for a single wall, or a distance greater than four times the effective opening for two intersecting walls.

[2] Vertical walls, ribs or similar obstructions extending from the water surface to or above the horizontal plane of the spout opening other than specified in Note 1 above. The effect of three or more such vertical walls or ribs has not been determined. In such cases, the airgap shall be measured from the top of the wall.

[3] The effective opening shall be the minimum cross-sectional area at the seat of the control valve or the supply pipe or tubing which feeds the device or outlet. If two or more lines supply one outlet, the effective opening shall be the sum of the cross-sectional areas of the individual supply lines or the area of the single outlet, whichever is smaller.

[4] Airgaps less than one (1) inch (25.4 mm) shall only be approved as a permanent part of a listed assembly that has been tested under actual backflow conditions with vacuums of 0 to 25 inches (635 mm) of mercury.

FIGURE 13.32 ■ Minimum airgaps for water distribution. (*Courtesy of International Code Council, Inc. and the International Plumbing Code 2000*)

Minimum Required Air Chamber Dimensions					
Nominal Pipe Diameter	Length of Pipe (ft.)	Flow Pressure P.S.I.G.	Velocity In Ft. Per. Sec.	Required Vol. In Cubic Inch	Air Chamber Phys. Size In Inches
1/2" (15 mm)	25	30	10	8	3/4" x 15"
1/2" (15 mm)	100	60	10	60	1" x 69.5"
3/4" (20 mm)	50	60	5	13	1" x 5"
3/4" (20 mm)	200	30	10	108	1.25" x 72.5"
1" (25 mm)	100	60	5	19	1.25" x 12.7 "
1" (25 mm)	50	30	10	40	1.25" x 27"
1-1/4" (32 mm)	50	60	10	110	1.25" x 54"
1-1/2" (40 mm)	200	30	5	90	2" x 27"
1-1/2" (40 mm)	50	60	10	170	2" x 50.5"
2" (50 mm)	100	30	10	329	3" x 44.5"
2" (50 mm)	25	60	10	150	2.5" x 31"
2" (50 mm)	200	60	5	300	3" x 40.5"

FIGURE 13.33 ▪ Minimum required air chamber dimensions. (*Courtesy of International Code Council, Inc. and the International Plumbing Code 2000*)

STACK SIZES FOR BEDPAN STEAMERS
AND BOILING-TYPE STERILIZERS
(Number of Connections of Various Sizes
Permitted to Various-sized Sterilizer Vent Stacks)

STACK SIZE (Inches)	CONNECTION SIZE		
	1 1/2"		2"
1 1/2" a	1	or	0
2 a	2	or	1
2 b	1	and	1
3 a	4	or	2
3 b	2	and	2
4 a	8	or	4
4 b	4	and	4

For SI: 1 inch = 25.4 mm.
a. Total of each size.
b. Combination of sizes.

FIGURE 13.34 ▪ Stack sizes for bedpan steamers and boiling-type sterilizers. (*Courtesy of International Code Council, Inc. and the International Plumbing Code 2000*)

SPECIALTY PLUMBING

STACK SIZES FOR PRESSURE STERILIZERS
(Number of Connections of Various Sizes Permitted
to Various-sized Vent Stacks)

STACK SIZE (inches)	CONNECTION SIZE			
	3/4"	1"	1 1/4"	1 1/2"
1 1/2[a]	3 or	2 or	1	
1 1/2[b]	2 and	1		
2[a]	6 or	3 or	2 or	1
2[b]	3 and	2		
2[b]	2 and	1 and	1	
2[b]	1 and	1 and		1
3[a]	15 or	7 or	5 or	3
3[b]		1 and	2 and	2
	1 and	5 and		1

For SI: 1 inch = 25.4 mm.

a. Total of each size.

b. Combination of sizes.

FIGURE 13.35 ■ Stack sizes for pressure sterilizers. (*Courtesy of International Code Council, Inc. and the International Plumbing Code 2000*)

	Minimum Flow Rates
Oxygen	.71 CFM per outlet[1] (20 LPM)
Nitrous Oxide	.71 CFM per outlet[1] (20 LPM)
Medical Compressed Air	.71 CFM per outlet[1] (20 LPM)
Nitrogen	15 CFM (0.42 m³/min.) free air per outlet
Vacuum	1 SCFM (0.03 Sm³/min.) per inlet[2]
Carbon Dioxide	.71 CFM per outlet[1] (20 LPM)
Helium	.71 CFM per outlet (20 LPM)

[1] Any room designed for a permanently located respiratory ventilator or anesthesia machine shall have an outlet capable of a flow rate of 180 LPM (6.36 CFM) at the station outlet.

[2] For testing and certification purposes, individual station inlets shall be capable of a flow rate of 3 SCFM, while maintaining a system pressure of not less than 12 inches (305 mm) at the nearest adjacent vacuum inlet.

FIGURE 13.36 ■ Minimum flow rates. (*Courtesy of International Code Council, Inc. and the International Plumbing Code 2000*)

Location of Gray-Water System

Minimum Horizontal Distance in Clear Required From:	Holding Tank		Irrigation/ Disposal Field	
	Feet	(mm)	Feet	(mm)
Building Structures[1]	5[2]	(1524 mm)	2[3]	(610 mm)
Property line adjoining private property	5	(1524 mm)	5	(1524 mm)
Water supply wells[4]	50	(15240 mm)	100	(30480 mm)
Streams and lakes[4]	50	(15240 mm)	50[5]	(15240 mm)
Sewage pits or cesspools	5	(1524 mm)	5	(1524 mm)
Disposal field and 100% expansion area	5	(1524 mm)	4[6]	(1219 mm)
Septic tank	0	(0)	5	(1524 mm)
On-site domestic water service line	5	(1524 mm)	5	(1524 mm)
Pressurized public water main	10	(3048 mm)	10[7]	(3048 mm)

Notes: When irrigation/disposal fields are installed in sloping ground, the minimum horizontal distance between any part of the distribution system and the ground surface shall be fifteen (15) feet (4572 mm).

[1] Including porches and steps, whether covered or uncovered, breezeways, roofed porte-cocheres, roofed patios, carports, covered walks, covered driveways and similar structures or appurtenances.

[2] The distance may be reduced to zero feet for above ground tanks when first approved by the Administrative Authority.

[3] Assumes a 45 degree (0.79 rad) angle from foundation.

[4] Where special hazards are involved, the distance required shall be increased as may be directed by the Administrative Authority.

[5] These minimum clear horizontal distances shall also apply between the irrigation/disposal field and the ocean mean higher high tide line.

[6] Plus two (2) feet (610 mm) for each additional foot of depth in excess of one (1) foot (305 mm) below the bottom of the drain line.

[7] For parallel construction/for crossings, approval by the Administrative Authority shall be required.

FIGURE 13.37 ▪ Location of gray water system. (*Courtesy of International Code Council, Inc. and the International Plumbing Code 2000*)

GRAY WATER SYSTEMS

Design Criteria of Six Typical Soils		
Type of Soil	Minimum square feet of irrigation/leaching area per 100 gallons of estimated graywater discharge per day	Maximum absorption capacity in gallons per square foot of irrigation/leaching area for a 24-hour period
Coarse sand or gravel	20	5.0
Fine sand	25	4.0
Sandy loam	40	2.5
Sandy clay	60	1.7
Clay with considerable sand or gravel	90	1.1
Clay with small amounts of sand or gravel	120	0.8

FIGURE 13.38 ■ Design criteria of six typical soils. (*Courtesy of International Code Council, Inc. and the International Plumbing Code 2000*)

Design Criteria of Six Typical Soils		
Type of Soil	Minimum square meters of irrigation/leaching area per liter of estimated graywater discharge per day	Maximum absorption capacity in liters per square meter of irrigation/leaching area for a 24-hour period
Coarse sand or gravel	0.005	203.7
Fine sand	0.006	162.9
Sandy loam	0.010	101.8
Sandy clay	0.015	69.2
Clay with considerable sand or gravel	0.022	44.8
Clay with small amounts of sand or gravel	0.030	32.6

FIGURE 13.39 ■ Design criteria of six typical soils. (*Courtesy of International Code Council, Inc. and the International Plumbing Code 2000*)

RAINFALL RATES

RATES OF RAINFALL FOR VARIOUS CITIES

Rainfall rates, in inches per hour, are based on a storm of one-hour duration and a 100-year return period. The rainfall rates shown in the appendix are derived from Figure 1106.1.

Alabama:
Birmingham 3.8
Huntsville 3.6
Mobile 4.6
Montgomery 4.2

Alaska:
Fairbanks 1.0
Juneau 0.6

Arizona:
Flagstaff 2.4
Nogales 3.1
Phoenix 2.5
Yuma 1.6

Arkansas:
Fort Smith 3.6
Little Rock 3.7
Texarkana 3.8

California:
Barstow 1.4
Crescent City 1.5
Fresno 1.1
Los Angeles 2.1
Needles 1.6
Placerville 1.5
San Fernando 2.3
San Francisco 1.5
Yreka 1.4

Colorado:
Craig 1.5
Denver 2.4
Durango 1.8
Grand Junction ... 1.7
Lamar 3.0
Pueblo 2.5

Connecticut:
Hartford 2.7
New Haven 2.8
Putnam 2.6

Delaware:
Georgetown 3.0
Wilmington 3.1

District of Columbia:
Washington 3.2

Florida:
Jacksonville 4.3
Key West 4.3
Miami 4.7
Pensacola 4.6
Tampa 4.5

Georgia:
Atlanta 3.7
Dalton 3.4
Macon 3.9
Savannah 4.3
Thomasville 4.3

Hawaii:
Hilo 6.2
Honolulu 3.0
Wailuku 3.0

Idaho:
Boise 0.9
Lewiston 1.1
Pocatello 1.2

Illinois:
Cairo 3.3
Chicago 3.0
Peoria 3.3
Rockford 3.2
Springfield 3.3

Indiana:
Evansville 3.2
Fort Wayne 2.9
Indianapolis 3.1

Iowa:
Davenport 3.3
Des Moines 3.4
Dubuque 3.3
Sioux City 3.6

Kansas:
Atwood 3.3
Dodge City 3.3
Topeka 3.7
Wichita 3.7

Kentucky:
Ashland 3.0
Lexington 3.1
Louisville 3.2
Middlesboro 3.2
Paducah 3.3

Louisiana:
Alexandria 4.2
Lake Providence .. 4.0
New Orleans 4.8
Shreveport 3.9

Maine:
Bangor 2.2
Houlton 2.1
Portland 2.4

Maryland:
Baltimore 3.2
Hagerstown 2.8
Oakland 2.7
Salisbury 3.1

Massachusetts:
Boston 2.5
Pittsfield 2.8
Worcester 2.7

Michigan:
Alpena 2.5
Detroit 2.7
Grand Rapids ... 2.6
Lansing 2.8
Marquette 2.4
Sault Ste. Marie ... 2.2

Minnesota:
Duluth 2.8
Grand Marais ... 2.3
Minneapolis 3.1
Moorhead 3.2
Worthington ... 3.5

Mississippi:
Biloxi 4.7
Columbus 3.9
Corinth 3.6
Natchez 4.4
Vicksburg 4.1

Missouri:
Columbia 3.2
Kansas City 3.6
Springfield 3.4
St. Louis 3.2

Montana:
Ekalaka 2.5
Havre 1.6
Helena 1.5
Kalispell 1.2
Missoula 1.3

Nebraska:
North Platte 3.3
Omaha 3.8
Scottsbluff 3.1
Valentine 3.2

Nevada:
Elko 1.0
Ely 1.1
Las Vegas 1.4
Reno 1.1

New Hampshire:
Berlin 2.5
Concord 2.5
Keene 2.4

New Jersey:
Atlantic City 2.9
Newark 3.1
Trenton 3.1

New Mexico:
Albuquerque 2.0
Hobbs 3.0
Raton 2.5
Roswell 2.6
Silver City 1.9

New York:
Albany 2.5
Binghamton 2.3
Buffalo 2.3
Kingston 2.7
New York 3.0
Rochester 2.2

North Carolina:
Asheville 4.1
Charlotte 3.7
Greensboro 3.4
Wilmington 4.2

North Dakota:
Bismarck 2.8
Devils Lake 2.9
Fargo 3.1
Williston 2.6

Ohio:
Cincinnati 2.9
Cleveland 2.6
Columbus 2.8
Toledo 2.8

Oklahoma:
Altus 3.7
Boise City 3.3
Durant 3.8
Oklahoma City ... 3.8

Oregon:
Baker 0.9
Coos Bay 1.5
Eugene 1.3
Portland 1.2

Pennsylvania:
Erie 2.6
Harrisburg 2.8
Philadelphia 3.1
Pittsburgh 2.6
Scranton 2.7

FIGURE 13.40 ▪ Rates of rainfall. (*Courtesy of International Code Council, Inc. and the International Plumbing Code 2000*)

Rhode Island:		Utah:		Wisconsin:	
Block Island	2.75	Brigham City	1.2	Ashland	2.5
Providence	2.6	Roosevelt	1.3	Eau Claire	2.9
		Salt Lake City	1.3	Green Bay	2.6
South Carolina:		St. George	1.7	La Crosse	3.1
Charleston	4.3			Madison	3.0
Columbia	4.0			Milwaukee	3.0
Greenville	4.1	Vermont:			
		Barre	2.3	Wyoming:	
South Dakota:		Brattleboro	2.7	Cheyenne	2.2
Buffalo	2.8	Burlington	2.1	Fort Bridger	1.3
Huron	3.3	Rutland	2.5	Lander	1.5
Pierre	3.1			New Castle	2.5
Rapid City	2.9			Sheridan	1.7
Yankton	3.6	Virginia:		Yellowstone Park	1.4
		Bristol	2.7		
Tennessee:		Charlottesville	2.8		
Chattanooga	3.5	Lynchburg	3.2		
Knoxville	3.2	Norfolk	3.4		
Memphis	3.7	Richmond	3.3		
Nashville	3.3				
Texas:					
Abilene	3.6	Washington:			
Amarillo	3.5	Omak	1.1		
Brownsville	4.5	Port Angeles	1.1		
Dallas	4.0	Seattle	1.4		
Del Rio	4.0	Spokane	1.0		
El Paso	2.3	Yakima	1.1		
Houston	4.6				
Lubbock	3.3				
Odessa	3.2	West Virginia:			
Pecos	3.0	Charleston	2.8		
San Antonio	4.2	Morgantown	2.7		

For SI: 1 inch =25.4 mm.

Source: National Weather Service, National Oceanic and Atmospheric Administration, Washington, D.C.

FIGURE 13.40 ■ *(Continued)* Rates of rainfall. *(Courtesy of International Code Council, Inc. and the International Plumbing Code 2000)*

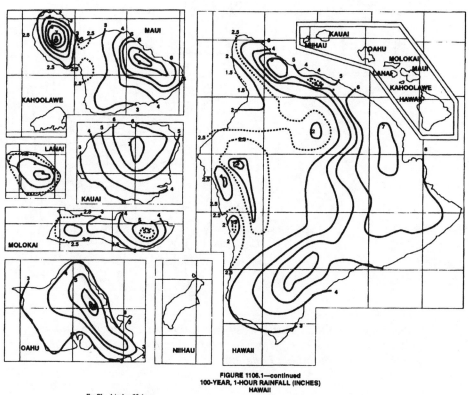

FIGURE 1106.1—continued
100-YEAR, 1-HOUR RAINFALL (INCHES)
HAWAII

For SI: 1 inch = 25.4 mm.
Source: National Weather Service, National Oceanic and Atmospheric Administration, Washington, DC.

FIGURE 13.41 ■ Hawaii figures show a 100-year, one-hour rainfall rate. (*Courtesy of International Code Council, Inc. and the International Plumbing Code 2000*)

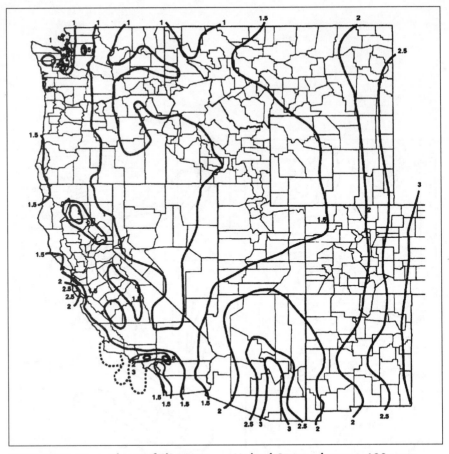

FIGURE 13.42 ■ Chart of the western United States shows a 100-year, one-hour rainfall rate. (*Courtesy of International Code Council, Inc. and the International Plumbing Code 2000*)

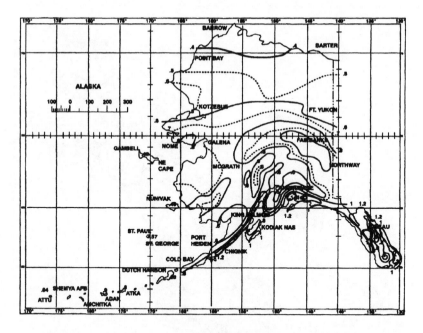

**100-YEAR, 1-HOUR RAINFALL (INCHES)
ALASKA**

For SI: 1 inch = 25.4 mm.

Source: National Weather Service, National Oceanic and Atmospheric Administration, Washington, DC.

FIGURE 13.43 ■ Alaska's 100-year, one-hour rainfall rate.
(*Courtesy of International Code Council, Inc. and the
International Plumbing Code 2000*)

FIGURE 13.44 ■ 100-year, one-hour rainfall rate for the eastern United States. (*Courtesy of International Code Council, Inc. and the International Plumbing Code 2000*)

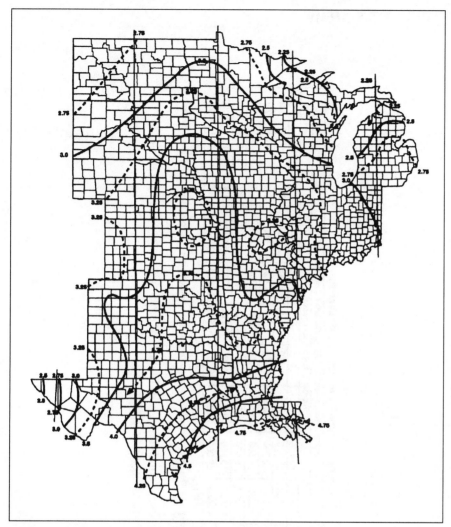

FIGURE 13.45 ■ 100-year, one-hour rainfall rate for the central United States. (*Courtesy of International Code Council, Inc. and the International Plumbing Code 2000*)

RAINWATER SYSTEMS

SIZE OF VERTICAL CONDUCTORS AND LEADERS

DIAMETER OF LEADER (inches)[a]	HORIZONTALLY PROJECTED ROOF AREA (square feet)											
	Rainfall rate (inches per hour)											
	1	2	3	4	5	6	7	8	9	10	11	12
2	2,880	1,440	960	720	575	480	410	360	320	290	260	240
3	8,800	4,400	2,930	2,200	1,760	1,470	1,260	1,100	980	880	800	730
4	18,400	9,200	6,130	4,600	3,680	3,070	2,630	2,300	2,045	1,840	1,675	1,530
5	34,600	17,300	11,530	8,650	6,920	5,765	4,945	4,325	3,845	3,460	3,145	2,880
6	54,000	27,000	17,995	13,500	10,800	9,000	7,715	6,750	6,000	5,400	4,910	4,500
8	116,000	58,000	38,660	29,000	23,200	19,315	16,570	14,500	12,890	11,600	10,545	9,660

For SI: 1 inch = 25.4 mm, 1 square foot = 0.0929 m².

a. Sizes indicated are the diameter of circular piping. This table is applicable to piping of other shapes provided the cross-sectional shape fully encloses a circle of the diameter indicated in this table.

FIGURE 13.46 ▪ Size of vertical conductors and leaders. (*Courtesy of International Code Council, Inc. and the International Plumbing Code 2000*)

SIZE OF HORIZONTAL STORM DRAINAGE PIPING

SIZE OF HORIZONTAL PIPING (inches)	HORIZONTALLY PROJECTED ROOF AREA (square feet)					
	Rainfall rate (inches per hour)					
	1	2	3	4	5	6
	1/8 unit vertical in 12 units horizontal (1-percent slope)					
3	3,288	1,644	1,096	822	657	548
4	7,520	3,760	2,506	1,800	1,504	1,253
5	13,360	6,680	4,453	3,340	2,672	2,227
6	21,400	10,700	7,133	5,350	4,280	3,566
8	46,000	23,000	15,330	11,500	9,200	7,600
10	82,800	41,400	27,600	20,700	16,580	13,800
12	133,200	66,600	44,400	33,300	26,650	22,200
15	218,000	109,000	72,800	59,500	47,600	39,650
	1/4 unit vertical in 12 units horizontal (2-percent slope)					
3	4,640	2,320	1,546	1,160	928	773
4	10,600	5,300	3,533	2,650	2,120	1,766
5	18,880	9,440	6,293	4,720	3,776	3,146
6	30,200	15,100	10,066	7,550	6,040	5,033
8	65,200	32,600	21,733	16,300	13,040	10,866
10	116,800	58,400	38,950	29,200	23,350	19,450
12	188,000	94,000	62,600	47,000	37,600	31,350
15	336,000	168,000	112,000	84,000	67,250	56,000
	1/2 unit vertical in 12 units horizontal (4-percent slope)					
3	6,576	3,288	2,295	1,644	1,310	1,096
4	15,040	7,520	5,010	3,760	3,010	2,500
5	26,720	13,360	8,900	6,680	5,320	4,450
6	42,800	21,400	13,700	10,700	8,580	7,140
8	92,000	46,000	30,650	23,000	18,400	15,320
10	171,600	85,800	55,200	41,400	33,150	27,600
12	266,400	133,200	88,800	66,600	53,200	44,400
15	476,000	238,000	158,800	119,000	95,300	79,250

For SI: 1 inch = 25.4 mm, 1 square foot = 0.0929 m².

FIGURE 13.47 ▪ Size of horizontal storm drainage piping. *(Courtesy of International Code Council, Inc. and the International Plumbing Code 2000)*

Sizing of Horizontal Rainwater Piping

Size of Pipe, Inches	Flow at 1/8"/ft. Slope, gpm	Maximum Allowable Horizontal Projected Roof Areas Square Feet at Various Rainfall Rates					
		1"/hr	2"/hr	3"/hr	4"/hr	5"/hr	6"/hr
3	34	3288	1644	1096	822	657	548
4	78	7520	3760	2506	1880	1504	1253
5	139	13,360	6680	4453	3340	2672	2227
6	222	21,400	10,700	7133	5350	4280	3566
8	478	46,000	23,000	15,330	11,500	9200	7670
10	860	82,800	41,400	27,600	20,700	16,580	13,800
12	1384	133,200	66,600	44,400	33,300	26,650	22,200
15	2473	238,000	119,000	79,333	59,500	47,600	39,650

Size of Pipe, Inches	Flow at 1/4"/ft. Slope, gpm	Maximum Allowable Horizontal Projected Roof Areas Square Feet at Various Rainfall Rates					
		1"/hr	2"/hr	3"/hr	4"/hr	5"/hr	6"/hr
3	48	4640	2320	1546	1160	928	773
4	110	10,600	5300	3533	2650	2120	1766
5	196	18,880	9440	6293	4720	3776	3146
6	314	30,200	15,100	10,066	7550	6040	5033
8	677	65,200	32,600	21,733	16,300	13,040	10,866
10	1214	116,800	58,400	38,950	29,200	23,350	19,450
12	1953	188,000	94,000	62,600	47,000	37,600	31,350
15	3491	336,000	168,000	112,000	84,000	67,250	56,000

Size of Pipe, Inches	Flow at 1/2"/ft. Slope, gpm	Maximum Allowable Horizontal Projected Roof Areas Square Feet at Various Rainfall Rates					
		1"/hr	2"/hr	3"/hr	4"/hr	5"/hr	6"/hr
3	68	6576	3288	2192	1644	1310	1096
4	156	15,040	7520	5010	3760	3010	2500
5	278	26,720	13,360	8900	6680	5320	4450
6	445	42,800	21,400	14,267	10,700	8580	7140
8	956	92,000	46,000	30,650	23,000	18,400	15,320
10	1721	165,600	82,800	55,200	41,400	33,150	27,600
12	2768	266,400	133,200	88,800	66,600	53,200	44,400
15	4946	476,000	238,000	158,700	119,000	95,200	79,300

Notes:
1. The sizing data for horizontal piping is based on the pipes flowing full.
2. For rainfall rates other than those listed, determine the allowable roof area by dividing the area given in the 1 inch/hour (25 mm/hour) column by the desired rainfall rate.

FIGURE 13.48 ■ Sizing of horizontal rainwater piping. (*Courtesy of International Code Council, Inc. and the International Plumbing Code 2000*)

Size of Gutters

Diameter of Gutter in Inches 1/16"/ft. Slope	Maximum Rainfall in Inches per Hour 2	3	4	5	6
3	340	226	170	136	113
4	720	480	360	288	240
5	1250	834	625	500	416
6	1920	1280	960	768	640
7	2760	1840	1380	1100	918
8	3980	2655	1990	1590	1325
10	7200	4800	3600	2880	2400

Diameter of Gutter in Inches 1/8"/ft. Slope	Maximum Rainfall in Inches per Hour 2	3	4	5	6
3	480	320	240	192	160
4	1020	681	510	408	340
5	1760	1172	880	704	587
6	2720	1815	1360	1085	905
7	3900	2600	1950	1560	1300
8	5600	3740	2800	2240	1870
10	10,200	6800	5100	4080	3400

Diameter of Gutter in Inches 1/4"/ft. Slope	Maximum Rainfall in Inches per Hour 2	3	4	5	6
3	680	454	340	272	226
4	1440	960	720	576	480
5	2500	1668	1250	1000	834
6	3840	2560	1920	1536	1280
7	5520	3680	2760	2205	1840
8	7960	5310	3980	3180	2655
10	14,400	9600	7200	5750	4800

Diameter of Gutter in Inches 1/2"/ft. Slope	Maximum Rainfall in Inches per Hour 2	3	4	5	6
3	960	640	480	384	320
4	2040	1360	1020	816	680
5	3540	2360	1770	1415	1180
6	5540	3695	2770	2220	1850
7	7800	5200	3900	3120	2600
8	11,200	7460	5600	4480	3730
10	20,000	13,330	10,000	8000	6660

FIGURE 13.49 ■ Size of gutters. *(Courtesy of International Code Council, Inc. and the International Plumbing Code 2000)*

SIZE OF SEMICIRCULAR ROOF GUTTERS

DIAMETER OF GUTTERS (inches)	HORIZONTALLY PROJECTED ROOF AREA (square feet) RAINFALL RATE (inches per hour)					
	1	2	3	4	5	6
1/16 unit vertical in 12 units horizontal (0.5-percent slope)						
3	680	340	226	170	136	113
4	1,440	720	480	360	288	240
5	2,500	1,250	834	625	500	416
6	3,840	1,920	1,280	960	768	640
7	5,520	2,760	1,840	1,380	1,100	918
8	7,960	3,980	2,655	1,990	1,590	1,325
10	14,400	7,200	4,800	3,600	2,880	2,400
1/8 unit vertical in 12 units horizontal (1-percent slope)						
3	960	480	320	240	192	160
4	2,040	1,020	681	510	408	340
5	3,520	1,760	1,172	880	704	587
6	5,440	2,720	1,815	1,360	1,085	905
7	7,800	3,900	2,600	1,950	1,560	1,300
8	11,200	5,600	3,740	2,800	2,240	1,870
10	20,400	10,200	6,800	5,100	4,080	3,400
1/4 unit vertical in 12 units horizontal (2-percent slope)						
3	1,360	680	454	340	272	226
4	2,880	1,440	960	720	576	480
5	5,000	2,500	1,668	1,250	1,000	834
6	7,680	3,840	2,560	1,920	1,536	1,280
7	11,040	5,520	3,860	2,760	2,205	1,840
8	15,920	7,960	5,310	3,980	3,180	2,655
10	28,800	14,400	9,600	7,200	5,750	4,800
1/2 unit vertical in 12 units horizontal (4-percent slope)						
3	1,920	960	640	480	384	320
4	4,080	2,040	1,360	1,020	816	680
5	7,080	3,540	2,360	1,770	1,415	1,180
6	11,080	5,540	3,695	2,770	2,220	1,850
7	15,600	7,800	5,200	3,900	3,120	2,600
8	22,400	11,200	7,460	5,600	4,480	3,730
10	40,000	20,000	13,330	10,000	8,000	6,660

For SI: 1 inch = 25.4 mm, 1 square foot = 0.0929 m².

FIGURE 13.50 ■ Size of semicircular roof gutters. (*Courtesy of International Code Council, Inc. and the International Plumbing Code 2000*)

Controlled Flow Maximum Roof Water Depth

Roof Rise,* Inches	(mm)	Max Water Depth at Drain, Inches	(mm)
Flat	(Flat)	3	(76)
2	(51)	4	(102)
4	(102)	5	(127s)
6	(152)	6	(152)

*Vertical measurement from the roof surface at the drain to the highest point of the roof surface served by the drain, ignoring any local depression immediately adjacent to the drain.

FIGURE 13.51 ■ Controlled flow maximum roof water depth. (*Courtesy of International Code Council, Inc. and the International Plumbing Code 2000*)

The visual graphics here should serve you well in your career. Knowing and understanding your local code is very important, so spend time with your codebook to gain a complete understanding of your local codes. Keep in mind that the information in this chapter is based in the International Code. If you work with the Uniform code, you may discover some differences between local requirements and those shown here.

chapter 14

SEPTIC CONSIDERATIONS

Septic systems are common in rural housing locations. Many people who live outside the parameters of municipal sewers depend on septic systems to solve their sewage disposal problems. Plumbers who work in areas where private waste disposal systems are common often come into contact with problems associated with septic systems. Ironically, plumbers are rarely the right people to call for septic problems, but they are often the first group of people homeowners think of when experiencing septic trouble.

One reason that plumbers are called so frequently for septic problems is that the trouble appears to be a stopped-up drain. When a septic system is filled beyond capacity, backups occur in houses. Most homeowners call plumbers when this happens. Smart plumbers check the septic systems first and find out if they are at fault.

Backups in homes are not the only reason why plumbers need to know a little something about septic systems. Customers frequently have questions about their plumbing systems that can be influenced by a septic system. For example, is it all right to install a garbage disposer in a home that is served by a septic system. Some people think it is, and others believe it isn't. The answer to this question may not be left up to a plumber's personal opinion.

Considering all of the questions and concerns that customers might come to their plumbers with, I feel it is wise for plumbers to develop a general knowledge of septic systems. This chapter will help you achieve this goal. With that said, let me show you what is involved with septic systems.

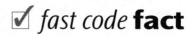

✓ *fast code* **fact**

Many local plumbing codes prohibit the installation of food grinders in homes where a septic system will receive the discharge.

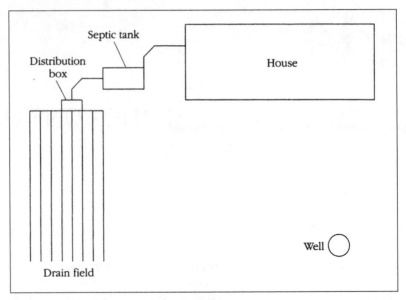

FIGURE 14.1 ■ A typical site plan.

SIMPLE SYSTEMS

Simple septic systems consist of a tank, some pipe, and some gravel. These systems are common, but they don't work well in all types of ground. Since most plumbers are not septic installers, I will not bore you will all of the sticky details for putting a pipe-and-gravel system into operation. However, I would like to give you a general overview of the system, so that you can talk intelligently with your customers.

THE COMPONENTS

Let's talk about the basic components of a pipe-and-gravel septic system. Starting near the foundation of a building, there is a sewer. The sewer pipe should be made of solid pipe, not perforated pipe. I know this seems obvious, but I did find a house a few years ago where the person who installed the sewer used perforated drain-field pipe. It was quite a mess. Most jobs today involve the use of schedule-40 plastic pipe for the sewer. Cast-iron pipe can be used, but plastic is the most common and is certainly acceptable.

The sewer pipe runs to the septic tank. There are many types of materials that septic tanks can be made of, but most of tanks are constructed of concrete. It is possible to build a septic tank on site, but every contractor I've ever known has bought pre-cast tanks. An average size tank holds about 1,000 gallons. The connection between the sewer and the septic tank should be watertight.

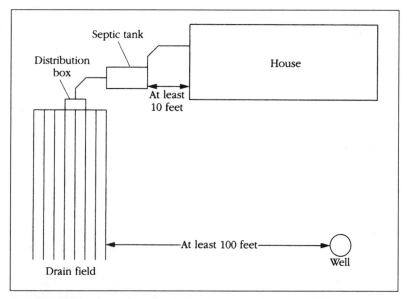

FIGURE 14.2 ■ Recommended minimum distances between wells and septic systems and septic tanks and homes.

The discharge pipe from the septic tank should be made of solid pipe, just like the sewer pipe. This pipe runs from the septic tank to a distribution box, which is also normally made of concrete. Once the discharge pipe reaches the distribution box, the type of materials used changes.

The drain field is constructed according to an approved septic design. In basic terms, the excavated area for the septic bed is lined with crushed stone. Perforated plastic pipe is installed in rows. The distance between the drainpipes and the number of drainpipes is controlled by the septic design. All of the drain-field pipes connect to the distribution box. The septic field is then covered with material specified in the septic design.

▶ *sensible* **shortcut**

Many contractors have been using fiberglass septic tanks in recent years. This can be a sensible solution to heavy, hard-to-handle concrete tanks.

☑ *fast code* **fact**

Many jurisdictions require septic designs to be drawn by certified, licensed designers.

As you can see, the list of materials is not a long one. Some schedule-40 plastic pipe, a septic tank, a distribution box, some crushed stone, and some perforated plastic pipe are the main ingredients. This is the primary reason why the cost of a pipe-and-gravel system is so low when compared to other types of systems.

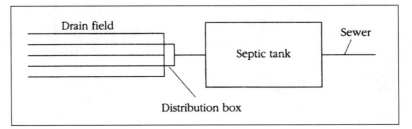

FIGURE 14.3 ■ Common septic layout.

Types Of Tanks

There are many types of septic tanks in use today. Pre-cast concrete tanks are, by far, the most common. However, they are not the only type of septic tank available. For this reason, let's discuss some of the material options that are available.

Pre-cast concrete is the most popular type of septic tank. When this type of tank is installed properly and is not abused, it can last almost indefinitely. However, heavy vehicular traffic running over the tank can damage it, so this situation should be avoided.

Metal septic tanks were once prolific. There are still a great number of them in use, but new installations rarely involve a metal tank. The reason is simple, metal tends to rust out, and that's not good for a septic tank. Some metal tanks are said to have given twenty years of good service. This may be true, but there are no guarantees that a metal tank will last even ten years. In all my years of being a contractor, I've never seen a metal septic tank installed. I've dug up old ones, but I've never seen a new one go in the ground.

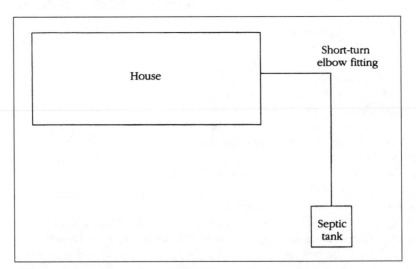

FIGURE 14.4 ■ Avoid using short-turn fittings between house and septic system.

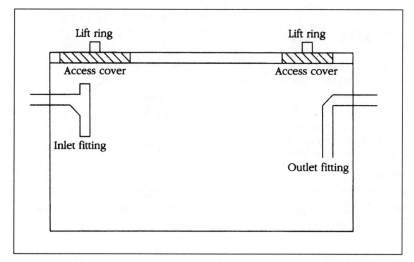

FIGURE 14.5 ■ Side view of a septic tank.

I don't have any personal experience with fiberglass septic tanks, but I can see some advantages to them. Their light weight is one nice benefit for anyone working to install the tank. Durability is another strong point in the favor of fiberglass tanks. However, I'm not sure how the tanks perform under the stress of being buried. I assume that their performance is good, but again, I have no first-hand experience with them.

Wood seems like a strange material to use for the construction of a septic tank, but I've read where it is used. The wood of choice, as I understand it,

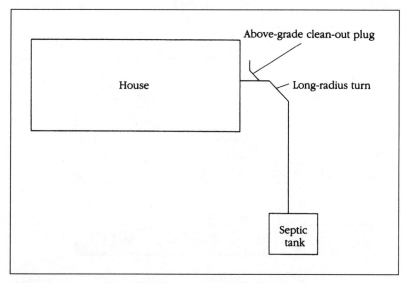

FIGURE 14.6 ■ Outside cleanout installed in sewer pipe and sweep-type fittings used to avoid pipe stoppages.

☞ been there **done that**

Some contractors turn to creative solutions to save money, but they may be making trouble for themselves. I prefer to use proven materials to avoid problems down the road. Compared to the cost of a pre-cast septic tank, building a tank on site doesn't make sense to me. I suggest using known products that are less likely to create warranty problems for you.

is redwood. I guess if you can make hot tubs and spas out of it, you can make a septic tank out of it. However, I don't think I would be anxious to warranty a septic tank made of wood.

Brick and block have also been used to form septic tanks. When these methods are employed, some type of parging and waterproofing must be done on the interior of the vessel. Personally, I would not feel very comfortable with this type of setup. This is, again, material that I have never worked with in the creation of a septic tank, so I can't give you much in the way of case histories.

CHAMBER SYSTEMS

Chamber septic systems are used most often when the perk rate on ground is low. Soil with a rapid absorption rate can support a standard, pipe-and-gravel septic system. Clay and other types of soil may not. When bedrock is close to the ground, surface chambers are often used.

▶ *sensible* **shortcut**

Septic chambers may be made of concrete or plastic. Many contractors prefer plastic chambers, since they are easier to work with.

What is a chamber system? A chamber system is installed very much like a pipe-and-gravel system, except for the use of chambers. The chambers might be made of concrete or plastic. Concrete chambers are naturally more expensive to install. Plastic chambers are shipped in halves and put together in the field.

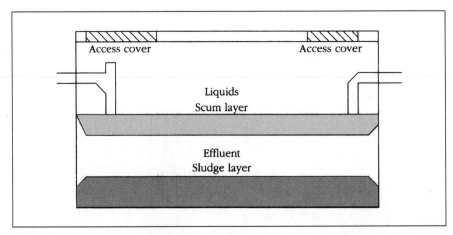

FIGURE 14.7 ■ Example of the various levels of materials in a septic system.

Since plastic is a very durable material, and it's relatively cheap, plastic chambers are more popular than concrete chambers.

When a chamber system is called for, there are typically many chambers involved. These chambers are installed in the leach field, between sections of pipe. As effluent is released from a septic tank, it is sent into the chambers. The chambers collect and hold the effluent for a period of time. Gradually, the liquid is released into the leach field and absorbed by the earth. The primary role of the chambers is to retard the distribution rate of the effluent.

Building a chamber system allows you to take advantage of land that would not be buildable with a standard pipe-and-gravel system. Based on this, chamber systems are good. However, when you look at the price tag of a chamber system, you may need a few moments to catch your breath. I've seen a number of quotes for these systems that pushed the $12,000 mark. This is more than double what the typical cost for a gravel-and-pipe system in my region. But, if you don't have any choice, what are you going to do?

A chamber system is simple enough in its design. Liquid leaves a septic tank and enters the first chamber. As more liquid is released from the septic tank, it is transferred into additional chambers that are farther downstream. This process continues with the chambers releasing a pre-determined amount of liquid into the soil as time goes on. The process allows more time for bacterial action to attack raw sewage, and it controls the flow of liquid into the ground.

If a perforated-pipe system was used in ground where a chamber system is recommended, the result could be a flooded leach field. This might create health risks. It would most likely produce unpleasant odors, and it might even shorten the life of the septic field.

> ### ▶ *sensible* **shortcut**

Contractors rarely make their own decisions on how to design or install a septic system. Rely on designs that are drawn by certified professionals. Don't cut corners to save a few dollars during the installation that could cost you major money when problems arise.

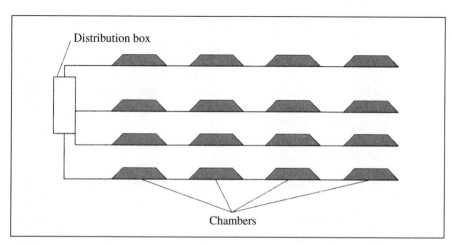

FIGURE 14.8 ■ Example of a chamber-type septic field.

Chambers are installed between sections of pipe within the drain field. The chambers are then covered with soil. The finished system is not visible above ground. All of the action takes place below grade. The only real downside to a chamber system is the cost.

TRENCH SYSTEMS

Trench systems are the least expensive versions of special septic systems. They are comparable in many ways to a standard pipe-and-gravel bed system. The main difference between a trench system and a bed system is that the drain lines in a trench system are separated by a physical barrier. Bed systems consist of drainpipes situated in a rock bed. All of the pipes are in one large bed. Trench fields depend on separation to work properly. To expand on this, let me give you some technical information.

A typical trench system is set into trenches that are between one to five feet deep. The width of the trench tends to run from one to three feet. Perforated pipe is placed in these trenches on a six-inch bed of crushed stone. A second layer of stone is placed on top of the drainpipe. This rock is covered with a barrier of some type to protect it from the backfilling process. The type of barrier used will be specified in a septic design.

☞ been there **done that**

When working with a trench system, there should be only one pipe in each trench. But, always follow the plans and specifications that have been approved by the local code officer.

When a trench system is used, both the sides of the trench and the bottom of the excavation are outlets for liquid. Only one pipe is placed in each trench. These two factors are what separate a trench system from a standard bed system. Bed systems have all of the drain pipes in one large excavation. In a bed system, the bottom of the bed is the only significant infiltrative surface. Since trench systems use both the bottoms and sides of trenches as infiltrative surfaces, more absorption is potentially possible.

▶ *sensible* **shortcut**

Because of their design, trench systems require more land area than bed systems do. This can be a problem on small building lots.

Neither bed nor trench systems should be used in soils where the percolation rate is either very fast or very slow. For example, if the soil will accept one inch of liquid per minute, it is too fast for a standard absorption system. This can be overcome by lining the infiltrative surface with a thick layer (about two feet or more) of sandy loam soil. Conversely, land that drains at a rate of one inch an hour is too slow for a bed or trench system. This is a situation where a chamber system might be recommended as an alternative.

Because of their design, trench systems require more land area than bed systems do. This can be a problem on small building lots. It can also add to the expense of clearing land for a septic field. However, trench systems are normally considered to be better than bed systems. There are many reasons for this.

Trench systems are said to offer up to five times more side area for infiltration to take place. This is based on a trench system with a bottom area identical to a bed system. The difference is in the depth and separation of the trenches. Experts like trench systems because digging equipment can straddle the trench locations during excavation. This reduces damage to the bottom soil and improves performance. In a bed system, equipment must operate within the bed, compacting soil and reducing efficiency.

If you are faced with hilly land to work with, a trench system is ideal. The trenches can be dug to follow the contour of the land. This gives you maximum utilization of the sloping ground. Infiltrative surfaces are maintained while excessive excavation is eliminated. The advantages of a trench system are numerous. For example, trenches can be run between trees. This reduces clearing costs and allows trees to remain for shade and

> ▶ *sensible* **shortcut**

The advantages of a trench system are numerous. For example, trenches can be run between trees. This reduces clearing costs and allows trees to remain for shade and aesthetic purposes. However, roots may still be a consideration.

aesthetic purposes. However, roots may still be a consideration. Most people agree that a trench system performs better than a bed system. When you combine performance with the many other advantages of a trench system, you may want to consider trenching your next septic system. It costs more to dig individual trenches than it does to create a group bed, but the benefits may outweigh the costs.

MOUND SYSTEMS

Mound systems, as you might suspect, are septic systems that are constructed in mounds that rise above the natural topography. This is done to compensate for high water tables and soils with slow absorption rates. Due to the amount of fill material to create a mound, the cost is naturally higher than it would be for a bed system.

Coarse gravel is normally used to build a septic mound. The stone is piled on top of the existing ground. However, topsoil is removed before the stone

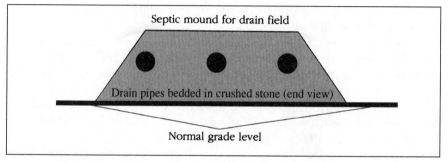

Septic mound for drain field

Drain pipes bedded in crushed stone (end view)

Normal grade level

FIGURE 14.9 ▪ Cut-away of a mount-type septic system.

is installed. When a mound is built, it contains suitable fill material, an absorption area, a distribution network, a cap, and topsoil. Due to the raised height, a mound system depends on either pumping or siphonic action to work properly. Essentially, effluent is either pumped or siphoned into the distribution network.

As the effluent is passing through the coarse gravel and infiltrating the fill material, treatment of the wastewater occurs. This continues as the liquid passes through the unsaturated zone of the natural soil. The purpose of the cap is to retard frost action, deflect precipitation, and to retain moisture that will stimulate the growth of ground cover.

> ☑ *fast code* **fact**
>
> Without adequate ground cover, erosion can be a problem. There are a multitude of choices available as acceptable ground covers. Grass is the most common choice.

Mounds should be used only in areas that drain well. The topography can be level or slightly sloping. The amount of slope allowable depends on the perk rate. For example, soil that perks at a rate of one inch every sixty minutes or less, should not have a slope of more than six percent if a mound system is to be installed. If the soil absorbs water from a perk test faster than one inch in one hour, the slope could be increased to twelve percent. These numbers are only examples. A professional who designs a mound system will set the true criteria for slope values. Ideally, about two feet of unsaturated soil should exist between the original soil surface and the seasonally saturated topsoil. There should be three to five feet of depth to the impermeable barrier. An overall range of perk rate could go as high as one inch in two hours, but this, of course, is subject to local approval. Perk tests for this type of system are best when done at a depth of about 20 inches. However, they can be performed at shallow depths of only 12 inches. Again, you must consult and follow local requirements.

HOW DOES A SEPTIC SYSTEM WORK?

How does a septic system work? A standard septic system works on a very simple principle. Sewage from a home enters the septic tank through the sewer. Where the sewer is connected to the septic tank, there is a baffle on the inside of the tank. This baffle is usually a sanitary tee. The sewer enters the center of the tee and drops down through the bottom of it. The top hub of the tee is left open.

The bottom of the tee is normally fitted with a short piece of pipe. The pipe drops out of the tee and extends into the tank liquids. This pipe should never extend lower than the outlet pipe at the other end of the septic tank. The inlet drop is usually no more than twelve inches long.

The outlet pipe for the tank also has a baffle, normally an elbow fitting. The drop from this baffle is frequently about sixteen inches in length.

When sewage enters a septic tank, the solids sink to the bottom of the tank and the liquids float within the confines of the container. As the tank collects waste, several processes begin to take place.

Solid waste that sinks to the bottom becomes what is known as sludge. Liquids, or effluent as it is called, are suspended between the lower layer of sludge and an upper layer of scum. The scum layer consists of solids and gases floating on the effluent. All three of these layers are needed for the waste disposal system to function properly.

As the effluent level rises in a tank, it eventually flows out of the tank, through the outlet pipe. The effluent drains down a solid pipe to the distribution box. After entering the distribution box, the liquid is routed into different slotted pipes that run through the leach field.

As the effluent mixes with air, aerobic bacteria begins to work on the waste. This bacteria attacks the effluent and eventually renders it harmless. Aerobic bacteria need oxygen to do their job. Drain fields should be constructed of porous soil or crushed stone to ensure the proper breakdown of the effluent.

As the effluent works its way through the drain field, it becomes odorless and harmless. By the time the effluent passes through the earth and becomes ground water, it should be safe to drink.

> ▶ *sensible* **shortcut**
>
> Anaerobic bacteria work inside the septic tank to break down the solids. This type of bacteria is capable of working in confinements void of oxygen. As the solids break up, they form the sludge layer.

> ▶ *sensible* **shortcut**
>
> Aerobic bacteria need oxygen to do their job.

SEPTIC TANK MAINTENANCE

Septic tank maintenance is not a time consuming process. Most septic systems require no attention for years at a time. However, when the scum and sludge layers combined have a depth of eighteen inches, the tank should be cleaned out.

Trucks equipped with suction hoses are normally used to clean septic tanks. The contents removed from septic tanks can be infested with germs. The disease risk of exposure to sludge requires that the sludge be handled carefully and properly.

HOW CAN CLOGS BE AVOIDED?

How can clogs be avoided? Clogs can be avoided by careful attention to what types of waste enter the septic system. Grease, for example, can cause a septic system to

> ☞ been there **done that**
>
> What happens if the drain field doesn't work? When a septic field fails to do its job, a health hazard exists. This situation demands immediate attention. The main reason for a field to fail in its operation is clogging. If the pipes in a drain field become clogged, they must be excavated and cleaned or replaced. If the field itself clogs, the leach bed must be cleaned or removed and replaced. Neither of these propositions is cheap.

become clogged. Bacteria does not do a good job in breaking down grease. Therefore, the grease can enter the slotted drains and leach field with enough bulk to clog them.

☑ *fast code* **fact**

Check your local code to see if garbage disposers can be installed in homes that are served by septic systems. Many jurisdictions do not allow garbage disposers in homes that depend on private sewage disposal systems.

Paper, other than toilet paper, can also clog up a septic system. If the paper is not broken down before entering the drain field, it can plug up the works.

WHAT ABOUT GARBAGE DISPOSERS, DO THEY HURT A SEPTIC SYSTEM?

What about garbage disposers, do they hurt a septic system? The answers offered to this question vary from yes to maybe to no. Many people, including numerous code enforcement offices, believe garbage disposers should not be used in conjunction with septic systems. Other people disagree and believe that disposers have no adverse effect on a septic system.

It is possible for the waste of a disposer to make it into the distribution pipes and drain field. If this happens, the risk for clogging is elevated. Another argument against disposers is the increased load of solids they put on a septic tank. Obviously, the amount of solid waste will depend on the frequency with which the disposer is used.

What is my opinion? My opinion is that disposers increase the risk of septic system failure and should not be used with such systems. However, I know of many houses using disposers with septic systems that are not experiencing any problems. If you check with your local plumbing inspector this question may become a moot point. Many local plumbing codes prevent the use of disposers with septic systems.

PIPING CONSIDERATIONS

There are some additional piping considerations for plumbers to observe. Septic tanks are designed to handle routine sewage. They are not meant to modify chemical discharges and high volumes of water. If, as a plumber, you pipe the discharge from a sump pump into the sanitary plumbing system, which you are not supposed to do, the increased volume of water in the tank could disrupt its normal operation.

GAS CONCENTRATIONS

👉 been there **done that**

Chemical drain openers used in high quantities can also destroy the natural order of a septic tank. Chemicals from photography labs are another risk plumbers should be aware of when piping drainage to a septic system.

Gas concentrations in a septic tank can cause problems for plumbers. The gases collected in a septic tank have the potential to explode. If you remove the top of a septic tank with a flame close by, you might be blown up. Also, breathing the gases for an extended period of time can cause health problems.

SEWAGE PUMPS

There are times when sewage pumps must be used to get sewage to a septic system. The pumps are normally installed in a buried box outside of the building being served. The box is often made of concrete. In these cases, the home's sewer pipe goes to the pumping station. From the pumping station, a solid pipe transports the waste to the septic tank.

Sewage pumps have floats that are lifted as the level of contents in the pump station build. When the float is raised to a certain point, the pump cuts on, emptying the contents of the pump station. The discharge pipe from the pump must be equipped with a check valve. Otherwise, gravity would force waste down the pipe, back into the pump station, when the pump cut off. This would result in the pump having to constantly cut on and off, wearing out the pump.

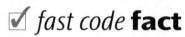

☑ *fast code* **fact**

Exterior sewage pumps must be equipped with alarm systems. The alarms warn the property owner if the pump is not operating and the pump station is filling with sewage. Without the alarm, the sewage could build to a point where it would flow back into the building.

☞ been there **done that**

Some homeowners associate an overflowing toilet with a problem in their septic system. It is possible that the septic system is responsible for the toilet backing up, but this is not always the case. A stoppage either in the toilet trap or in the drainpipe can cause a backup.

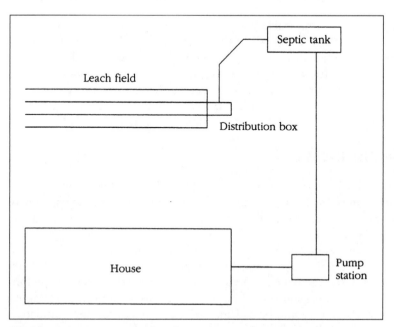

FIGURE 14.10 ■ Example of a pump-station septic system.

AN OVERFLOWING TOILET

Some homeowners associate an overflowing toilet with a problem in their septic system. It is possible that the septic system is responsible for the toilet backing up, but this is not always the case. A stoppage either in the toilet trap or in the drainpipe can cause a backup.

If you get a call from a customer who has a toilet flooding their bathroom, there is a quick, simple test you can have the homeowner perform to tell you more about the problem. You know the toilet drain is stopped up, but will the kitchen sink drain properly? Will other toilets in the house drain? If other fixtures drain just fine, the problem is not with the septic tank.

There are some special instructions that you should give your customers prior to having them test other fixtures. First, it is best if they use fixtures that are not in the same bathroom with the plugged-up toilet. Lavatories and bathing units often share the same main drain that a toilet uses. Testing a lavatory that is near a stopped-up toilet can tell you if the toilet is the only fixture affected. It can, in fact, narrow the likelihood of the problem down to the toilet's trap. But, if the stoppage is some way down the drainpipe, it's conceivable that the entire bathroom group will be affected. It is also likely that if the septic tank is the problem, water will back up in a bathtub.

When an entire plumbing system is unable to drain, water will rise to the lowest fixture, which is usually a bathtub or shower. So, if there is no backup in a bathing unit, there probably isn't a problem with a septic tank. But, backups in bathing units can happen even when the major part of a plumbing system is working fine. A stoppage in a main drain could cause the liquids to back up into a bathing unit.

To determine if a total backup is being caused, have homeowners fill their kitchen sinks and then release all of the water at once. Get them to do this several times. A volume of water may be needed to expose a problem. Simply running the faucet for a short while might not show a problem with the kitchen drain. If the kitchen sink drains successfully after several attempts, it's highly unlikely that there is a problem with the septic tank. This would mean that you should call your plumber, not your septic installer.

WHOLE-HOUSE BACKUPS

Whole-house backups (where none of the plumbing fixtures drain) indicate either a problem in the building drain, the sewer, or the septic system. There is no way to know where the problem is until some investigative work is done. It's possible that the problem is associated with the septic tank, but you will have to pinpoint the location where trouble is occurring.

For all the plumbing in a house to back up, there must be some obstruction at a point in the drainage or septic system beyond where the last plumbing drain enters the system. Plumbing codes require clean-out plugs along drainage pipes. There should be a clean-out either just inside the foundation wall of a home or just outside the wall. This clean-out location and the access panel of a septic tank are the two places to begin a search for the problem.

If the access cover of the septic system is not buried too deeply, I would start there. But, if extensive digging would be required to expose the cover, I would start with the clean-out at the foundation, hopefully on the outside of the house. Remove the clean-out plug and snake the drain. This will normally clear the stoppage, but you may not know what caused the problem. Habitual stoppages point to a problem in the drainage piping or septic tank.

Removing the inspection cover from the inlet area of a septic tank can show you a lot. For example, you may see that the inlet pipe doesn't have a tee fitting on it and has been jammed into a tank baffle. This could obviously account for some stoppages. Cutting the pipe off and installing the diversion fitting will solve this problem.

Sometimes pipes sink in the ground after they are buried. Pipes sometimes become damaged when a trench is backfilled. If a pipe is broken or depressed during backfilling, there can be drainage problems. When a pipe sinks in uncompacted earth, the grade of the pipe is altered, and stoppages become more likely. You might be able to see some of these problems from the access hole over the inlet opening of a septic tank.

Once you remove the inspection cover of a septic tank, look at the inlet pipe. It should be coming into the tank with a slight downward pitch. If the pipe is pointing upward, it indicates improper grading and a probable cause for stoppages. If the inlet pipe either doesn't exist or is partially pulled out of the tank, there's a very good chance that you have found the cause of your backup.

 been there **done that**

If a pipe is hit with a heavy load of dirt during backfilling, it can be broken off or pulled out of position. This won't happen if the pipe is supported properly before backfilling, but someone may have cheated a little during the installation.

In the case of a new septic system, a total backup is most likely to be the result of some failure in the piping system between the house and the septic tank. If your problem is occurring during very cold weather, it is possible that the drain pipe has retained water in a low spot and that the water has since frozen. I've seen it happen several times in Maine with older homes.

Running a snake from the house to the septic tank will tell you if the problem is in the piping. This is assuming that the snake used is a pretty big one. Little snakes might slip past a blockage that is capable of causing a backup. An electric drain-cleaner with a full-size head is the best tool to use.

THE PROBLEM IS IN THE TANK

There are times, even with new systems, when the problem causing a whole-house backup is in the septic tank. These occasions are rare, but they do exist. When this is the case, the top of the septic tank must be uncovered. Some tanks, like the one at my house, are only a few inches beneath the surface. Other tanks can be buried several feet below the finished grade.

Once a septic tank is in full operation, it works on a balance basis. The inlet opening of a septic tank is slightly higher than the outlet opening. When water enters a working septic tank, an equal amount of effluent leaves the tank. This maintains the needed balance. But, if the outlet opening is blocked by an obstruction, water can't get out. This will cause a backup.

Strange things sometimes happen on construction sites, so don't rule out any possibilities. It may not seem logical that a relatively new septic tank could be full or clogged, but don't bet on it. I can give you all kinds of things to think about. Suppose a septic installer was using up old scraps of pipe for drops and short pieces, and one of the pieces had a plastic test cap glued into the end of it that was not noticed? This could certainly render the septic system inoperative once the liquid rose to a point where it would be attempting to enter the outlet drain. Could this really happen? I've seen the same type of situation happen with interior plumbing, so it could happen with the piping at a septic tank.

What else could block the outlet of a new septic tank? Maybe a piece of scrap wood found its way into the septic tank during construction and is now blocking the outlet. If the wood floated in the tank and became aligned with the outlet drop, pressure could hold it in place and create a blockage. The point is that almost anything could be happening in the outlet opening, so take a snake and see if it is clear.

If the outlet opening is free of obstructions, and all drainage to the septic tank has been ruled out as a potential problem, you must look further down the line. Expose the distribution box and check it. Run a snake from the tank to the box. If it comes through without a hitch, the problem is somewhere in the leach field. In many cases, a leach field problem will cause the distribution box to flood. So, if you have liquid come rushing of the distribution box, you should be alerted to a probable field problem.

PROBLEMS WITH A LEACH FIELD

Problems with a leach field are uncommon among new installations. Unless the field was poorly designed or installed improperly, there is very little reason why it should fail. However, extremely wet ground conditions, due to heavy or constant rains, could force a field to become saturated. If the field saturates with ground water, it cannot accept the effluent from a septic tank. This, in turn, causes backups in houses. When this is the case, the person who created the septic design should be looked to in terms of fault.

Older Fields

Older fields sometimes clog up and fail. Some drain fields become clogged with solids. Financially, this is a devastating discovery. A clogged field has to be dug up and replaced. Much of the crushed stone might be salvageable, but the pipe, the excavation, and whatever new stone is needed can cost thousands of dollars. The reasons for a problem of this nature are either a poor design, bad workmanship, or abuse.

If the septic tank installed for a system is too small, solids are likely to enter the drain field. An undersized tank could be the result of a poor septic design, or it could come about as a family grows and adds onto their home.

A tank that is adequate for two people may not be able to keep up with the usage seen when four people are involved. Unfortunately, finding out that a tank is too small often doesn't happen until the damage has already been done.

Why would a small septic tank create problems with a drain field? Septic tanks accept solids and liquids. Ideally, only liquids should leave the septic tank and enter the leach field. Bacterial action occurs in a septic tank to break down solids. If a tank is too small, there is not adequate time for the breakdown of solids to occur. Increased loads on a small tank can force solids down into the drain field. After this happens for a while, the solids plug up the drainage areas in the field. This is when digging and replacement is needed.

In terms of a septic tank, a pipe with a fast grade can cause solids to be stirred up and sent down the outlet pipe. When a four-inch wall of water dumps into a septic tank at a rapid rate, it can create quite a ripple effect. The force of the water might generate enough stir to float solids that should be sinking. If these solids find their way into a leach field, clogging is likely.

> ☞ been there **done that**
>
> Is there any such thing as having too much pitch on a drainpipe. Yes, there is. A pipe that is graded with too much pitch can cause several problems. In interior plumbing, a pipe with a fast pitch may allow water to race by without removing all the solids. A properly graded pipe floats the solids in the liquid as drainage occurs. If the water is allowed to rush out, leaving the solids behind, a stoppage will eventually occur.

We talked a little bit about garbage disposers earlier. When a disposer is used in conjunction with a septic system, there are more solids involved that what would exist without a disposer. This, where code allows, calls for a larger septic tank. Due to the increase in solids, a larger tank is needed for satisfactory operation and a reduction in the risk of a clogged field. I remind you again, some plumbing codes prohibit the use of garbage disposers where a septic system is present.

Other causes for field failures can be related to collapsed piping. This is not common with today's modern materials, but it is a fact of life with some old drain fields. Heavy vehicular traffic over a field can compress it and cause the field to fail. This is true even of modern fields. Saturation of a drain field will cause it to fail. This could be the result of seasonal water tables or prolonged use of a field that is giving up the ghost.

Septic tanks should have the solids pumped out of them on a regular basis. For a normal residential system, pumping once every two years should be adequate. Septic professionals can measure sludge levels and determine if pumping is needed. Failure to pump a system routinely can result in a build-up of solids that may invade and clog a leach field.

Normally, septic systems are not considered to be a plumber's problem. Once you establish that a customer's grief is coming from a failed septic system, you should be off the hook. Advise your customers to call septic professionals and go onto your next service call; you've earned your money.

appendix 1

NATIONAL RAINFALL STATISTICS

National rainfall statistics are needed for computing the requirements of storm water systems. The expected rainfall rates are needed to figure out systems for roof drains, storm sewers, and similar methods of controlling storm water drainage. Fortunately, the rainfall rates for major cities are listed in this chapter. Similar information can often be found in plumbing codebooks. You will also find rain maps in this chapter and some codebooks. You can't accomplish much with only the rainfall rates. Consider the following information as reference material that you can use at anytime to compute the needs for controlling storm water. (Figs. A1.1 to A1.5)

Location	Rain (in inches per hour)
Alabama	
Birmingham	3.8
Huntsville	3.6
Mobile	4.6
Montgomery	4.2
Alaska	
Fairbanks	1.0
Juneau	0.6
Arizona	
Flagstaff	2.4
Nogales	3.1
Phoenix	2.5
Yuma	1.6
Arkansas	
Fort Smith	3.6
Little Rock	3.7
Texarkana	3.8
California	
Barstow	1.4
Crescent City	1.5
Fresno	1.1
Los Angeles	2.1
Needles	1.6
Placerville	1.5
San Fernando	2.3
San Francisco	1.5
Yreka	1.4
Colorado	
Craig	1.5
Denver	2.4
Durango	1.8
Grand Junction	1.7
Lamar	3.0
Pueblo	2.5
Connecticut	
Hartford	2.7
New Haven	2.8
Putnam	2.6
Delaware	
Georgetown	3.0
Wilmington	3.1
District of Columbia	
Washington	3.2

FIGURE A1.1 ▪ Rainfall rates.

Location	Rain (in inches per hour)
Florida	
Jacksonville	4.3
Key West	4.3
Miami	4.7
Pensacola	4.6
Tampa	4.5
Georgia	
Atlanta	3.7
Dalton	3.4
Macon	3.9
Savannah	4.3
Thomasville	4.3
Hawaii	
Hilo	6.2
Honolulu	3.0
Wailuku	3.0
Idaho	
Boise	0.9
Lewiston	1.1
Pocatello	1.2
Illinois	
Cairo	3.3
Chicago	3.0
Peoria	3.3
Rockford	3.2
Springfield	3.3
Indiana	
Evansville	3.2
Fort Wayne	2.9
Indianapolis	3.1
Iowa	
Davenport	3.3
Des Moines	3.4
Dubuque	3.3
Sioux City	3.6
Kansas	
Atwood	3.3
Dodge City	3.3
Topeka	3.7
Wichita	3.7
Kentucky	
Ashland	3.0
Lexington	3.1
Louisville	3.2
Middlesboro	3.2
Paducah	3.3

FIGURE A1.1 ■ (*Continued*) Rainfall rates.

Location	Rain (in inches per hour)
Louisiana	
Alexandria	4.2
Lake Providence	4.0
New Orleans	4.8
Shreveport	3.9
Maine	
Bangor	2.2
Houlton	2.1
Portland	2.4
Maryland	
Baltimore	3.2
Hagerstown	2.8
Oakland	2.7
Salisbury	3.1
Massachusetts	
Boston	2.5
Pittsfield	2.8
Worcester	2.7
Michigan	
Alpena	2.5
Detroit	2.7
Lansing	2.8
Grand Rapids	2.6
Marquette	2.4
Sault Ste. Marie	2.2
Minnesota	
Duluth	2.8
Grand Marais	2.3
Minneapolis	3.1
Moorhead	3.2
Worthington	3.5
Mississippi	
Biloxi	4.7
Columbus	3.9
Corinth	3.6
Natchez	4.4
Vicksburg	4.1
Missouri	
Columbia	3.2
Kansas City	3.6
Springfield	3.4
St. Louis	3.2

FIGURE A1.1 ■ (*Continued*) Rainfall rates.

Location	Rain (in inches per hour)
Montana	
Ekalaka	2.5
Havre	1.6
Helena	1.5
Kalispell	1.2
Missoula	1.3
Nebraska	
North Platte	3.3
Omaha	3.8
Scottsbluff	3.1
Valentine	3.2
Nevada	
Elko	1.0
Ely	1.1
Las Vegas	1.4
Reno	1.1
New Hampshire	
Berlin	2.5
Concord	2.5
Keene	2.4
New Jersey	
Atlantic City	2.9
Newark	3.1
Trenton	3.1
New Mexico	
Albuquerque	2.0
Hobbs	3.0
Raton	2.5
Roswell	2.6
Silver City	1.9
New York	
Albany	2.5
Binghamton	2.3
Buffalo	2.3
Kingston	2.7
New York	3.0
Rochester	2.2
North Carolina	
Bismarck	2.8
Devils Lake	2.9
Fargo	3.1
Williston	2.6
Ohio	
Cincinnati	2.9
Cleveland	2.6
Columbus	2.8
Toledo	2.8

FIGURE A1.1 ■ (*Continued*) Rainfall rates.

Location	Rain (in inches per hour)
Oklahoma	
Altus	3.7
Boise City	3.3
Durant	3.8
Oklahoma City	3.8
Oregon	
Baker	0.9
Coos Bay	1.5
Eugene	1.3
Portland	1.2
Pennsylvania	
Erie	2.6
Harrisburg	2.8
Philadelphia	3.1
Pittsburgh	2.6
Scranton	2.7
Rhode Island	
Providence	2.6
South Carolina	
Charleston	4.3
Columbia	4.0
Greenville	4.1
South Dakota	
Buffalo	2.8
Huron	3.3
Pierre	3.1
Rapid City	2.9
Yankton	3.6
Tennessee	
Chattanooga	3.5
Knoxville	3.2
Memphis	3.7
Nashville	3.3
Texas	
Abilene	3.6
Amarillo	3.5
Brownsville	4.5
Dallas	4.0
Del Rio	4.0
El Paso	2.3
Houston	4.6
Lubbock	3.3
Odessa	3.2
Pecos	3.0
San Antonio	4.2

FIGURE A1.1 ■ (*Continued*) Rainfall rates.

Location	Rain (in inches per hour)
Utah	
Brigham City	1.2
Roosevelt	1.3
Salt Lake City	1.3
St. George	1.7
Vermont	
Barre	2.3
Bratteboro	2.7
Burlington	2.1
Rutland	2.5
Virginia	
Bristol	2.7
Charlottesville	2.8
Lynchburg	3.2
Norfolk	3.4
Richmond	3.3
Washington	
Omak	1.1
Port Angeles	1.1
Seattle	1.4
Spokane	1.0
Yakima	1.1
West Virginia	
Charleston	2.8
Morgantown	2.7
Wisconsin	
Ashland	2.5
Eau Claire	2.9
Green Bay	2.6
La Crosse	3.1
Madison	3.0
Milwaukee	3.0
Wyoming	
Cheyenne	2.2
Fort Bridger	1.3
Lander	1.5
New Castle	2.5
Sheridan	1.7
Yellowstone Park	1.4

FIGURE A1.1 ■ (*Continued*) Rainfall rates.

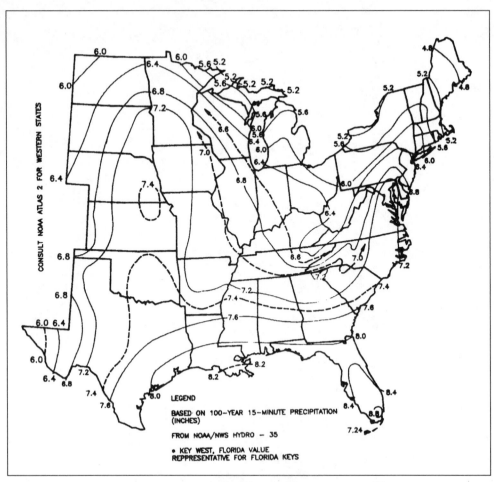

FIGURE A1.2 ■ Rainfall rates for secondary roof drains. (*Courtesy of McGraw-Hill*)

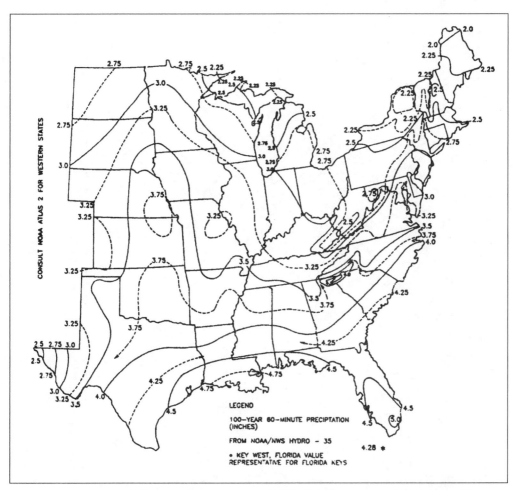

FIGURE A1.3 ■ Rainfall rates for primary roof drains. (*Courtesy of McGraw-Hill*)

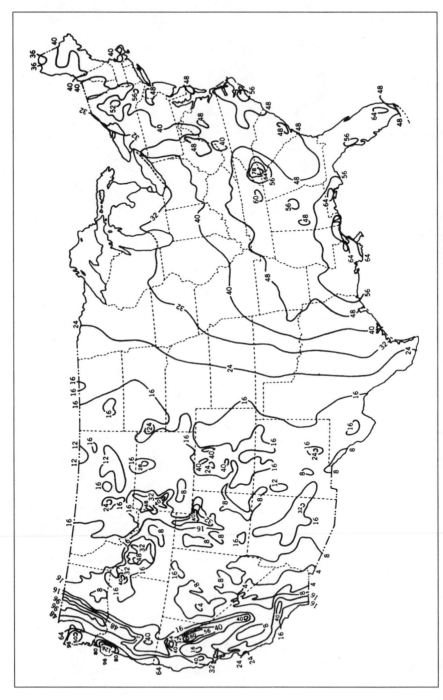

FIGURE A1.4 ■ Average annual precipitation in United States. (*Courtesy of McGraw-Hill*)

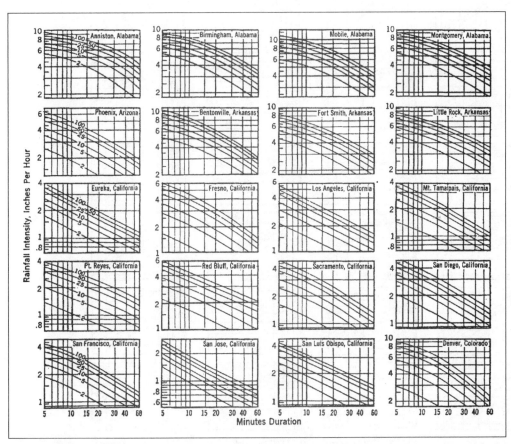

FIGURE A1.5 ■ Rainfall intensity-duration-frequency charts. (*Courtesy of McGraw-Hill*)

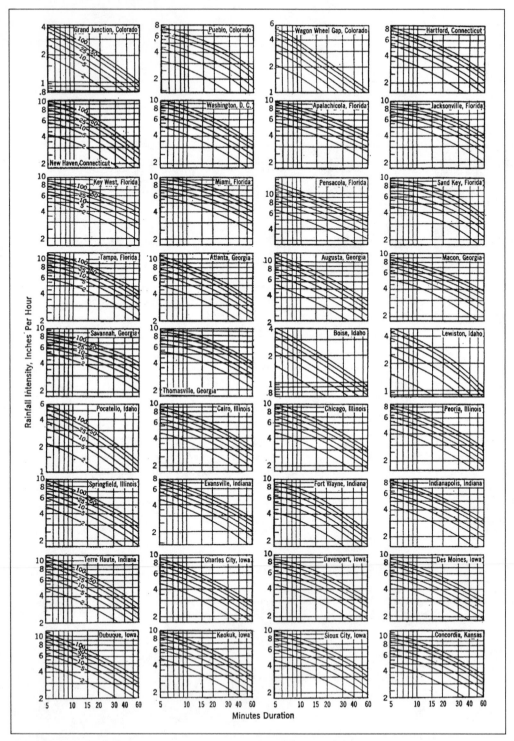

FIGURE A1.5 ■ (*Continued*) Rainfall intensity-duration-frequency charts. (*Courtesy of McGraw-Hill*)

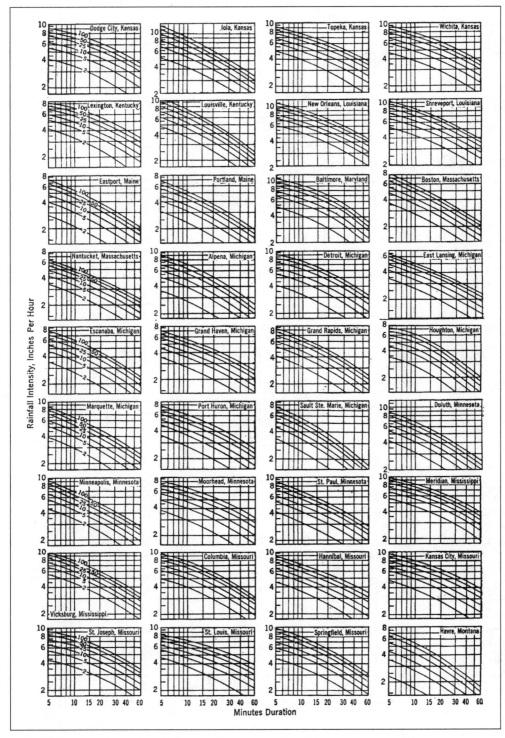

FIGURE A1.5 ■ (*Continued*) Rainfall intensity-duration-frequency charts. (*Courtesy of McGraw-Hill*)

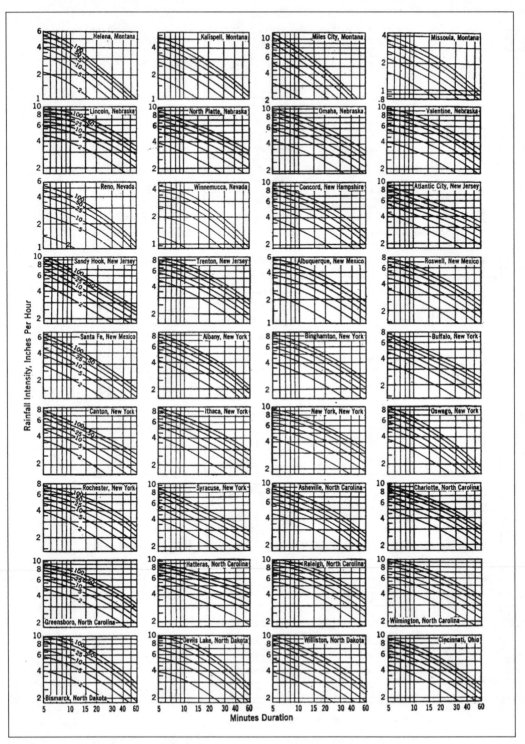

FIGURE A1.5 ■ (*Continued*) Rainfall intensity-duration-frequency charts. (*Courtesy of McGraw-Hill*)

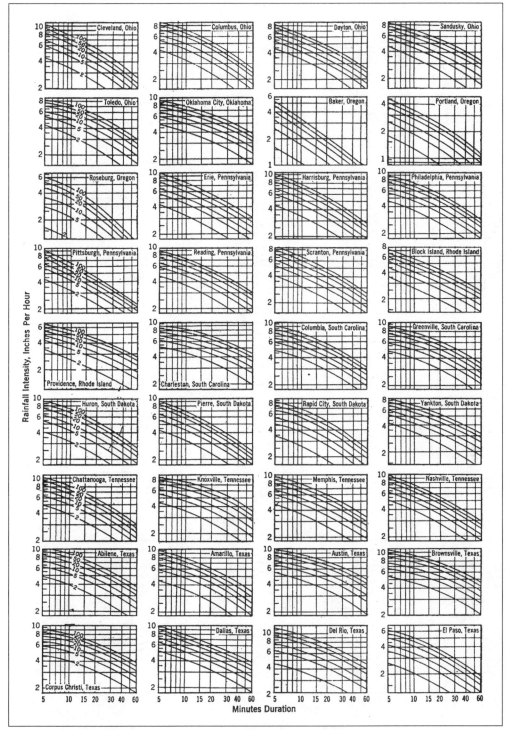

FIGURE A1.5 ▪ (*Continued*) Rainfall intensity-duration-frequency charts.
(*Courtesy of McGraw-Hill*)

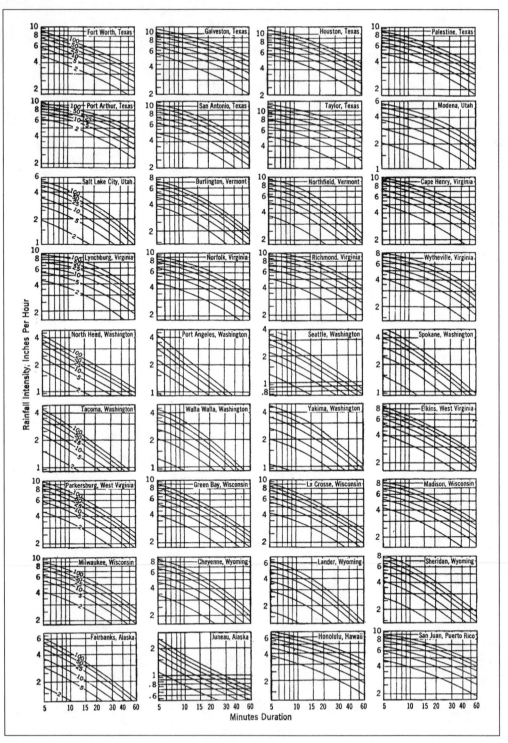

FIGURE A1.5 ■ (*Continued*) Rainfall intensity-duration-frequency charts. (*Courtesy of McGraw-Hill*)

INDEX

Page numbers in italics refer to figures and tables